Walking

the

Shores

of

Cape Cod

Walking
the
Shores
of
Cape Cod

Elliott Carr

With a Foreword by John Hay
Illustrations by Jane A. MacKenzie

Yarmouth Port

Third printing.

© Copyright 1997 by Elliott Carr
ISBN: 0-9653283-2-5

Illustrated by Jane A. MacKenzie.

Cover photo by Adam Gamble.
adam@oncapepublications.com

Cover and design by Joe Gallante, J Graphic Design
coysbrookstudio@comcast.net

All rights reserved.
No part of this book may be reproduced in any form,
including print and electronic means, without written permission
from On Cape Publications, with the exception
of brief quotations for review purposes.

Published by On Cape Publications
P.O. Box 218
Yarmouth Port, MA 02675
Toll free: 1-877-662-5839
Email: walkingcc@oncapepublications.com
On the web: www.oncapepublications.com

Printed in Canada.

Table of Contents

Dedication & Acknowledgements vii
Foreword ix
Preamble xiii
Coast Guard Beach to Race Point: The Outer Beach 1
Provincetown: Zoning 7
MacMillan Pier, Provincetown: Northern Right Whales 13
Truro: Disagreement & Due Process 19
Great Island, Wellfleet: History on the Shores of Cape Cod Bay .. 27
Campground Beach, Eastham: Erosion 35
Namskaket Marsh, Brewster: Salt Marshes & Wadlopen 45
The Sand Flats of Brewster: My "Home Beach" 49
Barnstable Harbor: Mixed Adventure 61
The Cape Cod Canal: Tides 67
Buzzards Bay: Family Walking 73
Patuisset Point, Bourne: Neighborhoods & Houses 77
Black Beach, Falmouth: Controlling Development 83
Woods Hole: Science & Economics 87
Falmouth Inner Harbor: Swimming Nantucket Sound 95
Waquoit Bay: Public Purchase of Land 99
Osterville: A Large House by the Sea 107
Wianno: Hurricanes 113
Craigville Beach: Public Beaches 121
Nantucket Sound: Trespassing 129
Monomoy Island, Chatham: Other Summer Visitors 135
Chatham Harbor: Man Against the Sea 145
Pleasant Bay: Powermen & Drifters 151
Town Cove, Orleans: Planning 159
Nauset Light, Eastham: Cussing the Moon 165

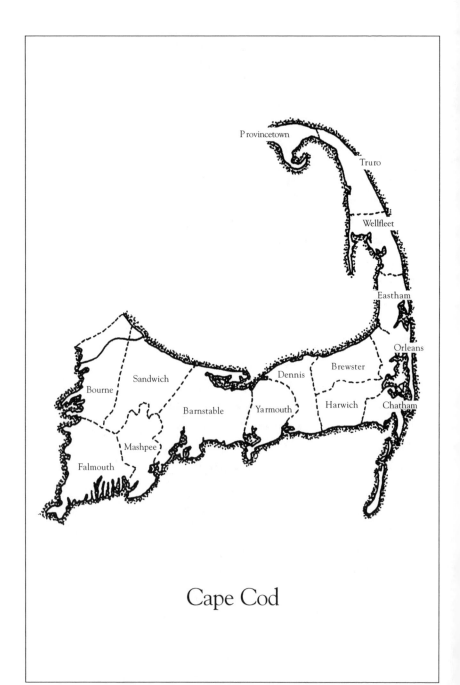

Cape Cod

Dedication & Acknowledgements

This book is dedicated to my wife Sue, my daughters Priscilla and Sarah, and all of the friends who walked with me, talked with me, and encouraged me to tell my story—Carol and Gordon Wright, Judy and Bob Wilkinson, Mykee and Bill Wibel, Bonnie and Meff Runyon, Janet and Peter Norton, Alicia and Bruce Hammatt, and Harry Terkanian are wonderful walking companions and guides. Andy Young took me out for a pleasant tour of Pleasant Bay. Larry Fox and Bob Finch read the original manuscript, probably barely recognizable in the final book. Victoria Ogden kept the project going by publishing a series of articles in The Cape Codder. Graham Giese, Peter Borrelli, Moira Brown, Noel Beyle, Hawkins Conrad, and Jack Pearce provided thoughtful input to portions of the final manuscript. Greg O'Brien's editorial advice was invaluable. Jane MacKenzie's drawings are better than my writing. Sue spent hours drawing maps. Without their assistance, and that of many others, I would never have completed the walk or finished telling the tale.

Foreword

In this book, Elliott Carr follows a number of predecessors who wrote about the Cape. One of these was Henry Kittredge of Barnstable. He was of a generation born before World War I, who were far closer to tradition and living memory than we can be today. Many of the people who could supply him with first hand source material about Cape Cod were still alive.

Other writers, as well as artists, both residents and itinerants, portrayed the Cape as a quiet secluded place of gentle hills and valleys, ponds and sand dunes and long beaches confronting the everlasting waters of the Atlantic. The Cape was relatively difficult to reach until after World War II.

Now you can get here in almost no time flat; and the traffic, which once slacked off after Labor Day, is unceasing, even in the wintertime. The Cape is so full of highways and "improved" roads that you can reach any part of it without getting out of your car. Houses are quickly built and easily exchangeable. And any remaining, open land fetches premium prices. The local speech is largely gone, and most of the small truck farmers and shore fishermen seem to have become an endangered species.

Elliott Carr, who is much interested in the right proportion between people and living space, has a new and welcome solution to life on the Cape. He walks, to find that walking is a far better way to measure place and time than speeding in an automobile. He and his wife Susan, sometimes accompanied by friends, have walked over 400 miles of the shoreline, from one end of the Cape to the other. They have hiked the Great Beach, estimating their number of footprints. They have walked across the wide flats off Dennis and Brewster, almost from dawn to dusk, or, at times, in the moonlight. This is too strenuous for most of the population, but it is the right way to encounter both open space and regional history.

This book recounts an incident in which they were picked up by the police, because they had been obliged to set foot on ground protected by zealous property owners. On several occasions Elliott was almost hopelessly mired in marsh mud, or "muck" as he prefers to call it. I suspect, at my age, that if I had attempted such muddy crossings I would now be a permanent part of the local material.

The author describes himself as a "populist" disclaiming any role as a naturalist, though his interest in tidal phenomena and whales, not to mention hurricanes, is keen. It is possible that the two terms are irreconcilable. But I suspect that they both meet in a nature we will not escape, commanded by powers of wind and tide we are unable to control.

Margaret Fuller, one of the Concord Transcendentalists, declared, in a romantic moment, that she had decided to "accept the universe." When this was repeated to her friend Ralph Waldo Emerson, he said: "She'd better."

As John Muir put it, going out is really going in. It is a voyage of self discovery. You can never find the real Cape until you leave the tyranny of a highway which imposes its own terms. You will never know fish without recognizing your dependence on them. Land itself cannot be understood simply on the basis of the limits put on it by real estate. The original past lies out along and beyond the shoreline, whose inhabitants have been working the tidelands for millions of years. The sea does not recognize temporary claims of ownership.

It is also true that walking across those far reaching tidelands is conducive to thought, even, or especially, when we are alone and can hear a greater distance in our minds.

I suspect, that as a fellow walker, it would be hard for me to catch up with Elliott, but walking can be the way to revelation.

John Hay
Brewster, Mass.

Preamble

One evening early in my walk around Cape Cod, my wife, Sue, and I decided to walk from Red River Beach in Harwich, around Morris Island in Chatham all the way to Chatham Harbor. The walk represented an excellent test: a test of maps and our ability to plan in an area unfamiliar to us, and a test of our ability to swim channels to shorten the route.

After a few miles of walking that included some major wading and some minor swimming to cross the Red River and the Cockle Cove area of Chatham, we came to the first major swim—the boat channel into Chatham's Stage Harbor. It appeared to be about a hundred yards across, just far enough to look like more than a casual swim after a considerable amount of walking. A sign prohibited swimming, which is never appropriately mixed with boating. Although it was a still evening, we had reached the channel slightly before low tide. The current continued to move out, making us wonder if we would be carried out into Nantucket Sound.

The only boat in sight was a slow moving sailboat. There was no turning back; we started to swim. About halfway across, Sue announced that her hip was getting sore. Her complaint, while not alarming enough to cause major concern, did make me wonder if we really knew what we were doing. I was much relieved as I touched bottom about two thirds of the way across.

Twilight fell, soon followed by darkness. We continued on in solitude through a cloud of "no-see-ems" and another swim across the channel in Chatham Harbor to reach our destination, Chatham Light.

My motivation to walk around Cape Cod evolved primarily from two character traits. I have always been easily seduced by the challenge of physical adventure and, as life progresses, have found it increasingly necessary to establish ever larger targets as a means of staying in shape. Combined, they have led me into a number of strenuous activities such as a one-day, 226-mile bicycle ride through all six New England states, a cross-country ski race from the top of Mount Mansfield in Vermont to the village of Stowe ten miles away, and numerous other shorter endurance events on bicycle, canoe, or foot. Frequently, in total exhaustion after some such adventure, I vowed "never to do anything so foolish again." But, when the next alluring adventure was announced, I would always again inevitably respond to some indefinable combination of the need to keep my middle aged body in shape, the need to believe I am younger than I am, the unparalleled satisfaction derived from total exhaustion, and—yes—the ability to say, "I did it."

I am not a naturalist or a historian; there is nothing about my background that indicates I would become interested in many of the attractions of Cape Cod which did come to absorb me as the walk progressed. I never studied or counted birds; my concern with tides was based solely on ascertaining convenient times to walk, and my interest in erosion was in how much land was being lost. Having grown up in New Hampshire, I came to Cape Cod two decades ago, not because of anything the Cape is, but because of something it is not—a city.

I spent some of my boyhood on the shore, a month each summer at my grandparents' cottage in Mattapoisett on the other side of Buzzard's Bay, and one full summer as a young boy in the early fifties at Camp Monomoy in Brewster when my parents wanted the summer off to enjoy themselves. But, to a young boy, the shoreline is a place you go to swim or boat—to have fun—certainly not to think. Forty-seven years later, my two strongest memories of Camp Monomoy are of watching planes bomb the target ship a mile away (yes, they really did) and watching the occasional car that passed by in the middle of the summer on Route 6A between Brewster and Orleans—just to see if it would turn in.

My walk around Cape Cod thus began solely as a physical challenge, a plan for organized walking. But it became much more. While walking, the pace is slow enough to enable one to see much more and to think in considerably more depth about what one is viewing. I first realized this on a walk with my friend, Bob Murray, who makes an annual trek from one end of Cape Cod to the other to raise funds for the homeless. One day we stopped for lunch at a new but unoccupied four-unit shopping strip on Route 28 in Chatham. After musing for a few moments about who might have been foolish enough to build it without tenants, I realized that the bank I work for owned the place, having foreclosed on it a few months before.

Walking around the Cape presents an excellent vantage point to view the collateral in both a specific and a general sense. What is this place in which we all live and do business? Cape Cod is, after all, little more than the coastline which I was traversing. If you took the entire Cape—without the shoreline—and plunked it down in the middle of Nebraska, it would do little to improve the charm or value of its new surroundings.

The perambulation thus soon became a learning experience, observing the intricately interwoven tapestry of the efforts of people to make a living from our coastal resources—the economy and the natural objects that make this place so attractive—the environment. The history of the interaction between the environment and the economy on Cape Cod has been direct and ever-changing. When different interests conflict, heated debate arises over issues such as rights of access to the coast, erosion and efforts to resist it, and the continued propriety of various forms of economic activity on our shores.

My journey was not a continuous point-to-point walk around the Cape. I walked where I could, when I could. After 40 walks of approximately 250 miles, broken up by 31 swims, I had completed a continuous route around the Cape's outer perimeter. The site and direction of each day's walk were determined by time, distance from home, and the availability of parking at the ends of the walks. Sue accompanied me for approximately half of the walks, initially in response to my pleas for her to park her car at one end or the other, but eventually because she, too, came to enjoy it. Despite the aforementioned story, Sue, unlike me, would rather swim than walk. The swims shortened the initial route by about 200 miles.

After completion of the original tour, I returned to walk many of the coves and rivers that were initially bypassed. This turned out to be the most difficult walking of the entire experience. In these estuaries there are more natural obstacles, such as salt marshes, than on the outer shore, and they are much more densely inhabited, requiring a walk there in the off-season. On Cape Cod, the bigger the house, the less likely it seems to be occupied year-round.

All told, the walk became a source of many pleasures:

- The physical challenge of swimming channels, slogging around salt marshes, and walking untold miles in loose sand.

- The joy of walking and talking with friends and the contemplative solitude of walking alone.

- The observation and study of natural objects, such as whales, horseshoe crabs, and ospreys; and of natural events, such as tides, hurricanes, and fog.

- The study of history as it had played itself out on our

shores including lighthouses, mooncussers, saltworks, and sailors.

• The observation of human conflicts as they arise and the efforts of our society to resolve them.

These phenomena that made the walk so interesting and enjoyable are all interrelated:

• The beauty of the natural environment is the heart of economic activity.

• The power of natural events, such as tides, storm, and fog, irrevocably influences all patterns of human activity.

• The beauty and mistakes of history sway current activity, often to repeat the past, but in other instances to produce the opposite result. The distinctive character of the people of Provincetown will probably never change, but these same people, who once brutally killed whales, now study them and protect them.

• The significant under-employed portion of the local population represents the greatest threat to the local environment.

• Ongoing destruction and privatization of portions of the local environment represent the greatest threat to the local economy.

To realize these pleasures, the walker of Cape Cod's shores must constantly cope with two problems.

The first is the need to traverse or circumvent private property. The Colonial Ordinances of 1641-47 granted ownership of the area between the high and low tide lines to property owners, except for the purposes of "fishing, fowling, and navigation." From a legal standpoint, there are thus three zones.

Wading below the low tide line is always legal, although if you try it in the type of neighborhood where there is a guard house, you may be stopped.

Walking between the high and low tide lines is only legal to fish or "fowl." Although the later term originally applied to bird hunting, Massachusetts Attorney General Scott Harshbarger recently defined "fowl" to include bird watching. He pointed out, however, that this interpretation has yet to be legally challenged. Some would thus say that it is legal to walk in this middle zone if you are carrying a fishing pole or a pair of binoculars, but I would not want to end up in court without bait and some plausible explanation of what type of fish I hoped to catch.

Walking above the high tide line is always on private property and therefore off bounds.

The second problem is related to the first. Almost all of our coast is easier to walk at low tide. Some sections are dangerous or impossible to walk on at high tide. I almost always walk the shore at low tide, even then constantly dressed and prepared to get my feet wet or muddy, or in many places even to do some serious wading or swimming. Such preparation expands the flexibility and the fun.

For most of the year a bathing suit and a pair of old sneakers over fast drying socks that wick away the water will suffice. In the winter, more impervious options continue to permit passage in many areas, but one must be more knowledgeable about local conditions—and risks.

I had a crisis about halfway through my walk around the Cape. The grubby old sneakers I had been wearing wore smooth and lost all traction on wet rocks. As I searched the closets for another suitable old pair, Sue presented me with the gift of new bridal white sneakers. I wore them, but immediately sought out some good muck to make me stop feeling like a nurse.

Walking The Shores of Cape Cod is not like most other books

on Cape Cod and is probably not what many readers will expect. Although there is much adventure in it, it is not an adventure book. Although there is much appreciation of natural beauty in it, it is not a book on the natural pleasures on Cape Cod's shores. Instead, it attempts to convey the immense variety of the experiences in which one can partake while walking these shores. By focusing on 25 very different places and very different subjects on the way around the Cape, the book is intended to be a salad of Cape Cod's pleasures and the problems facing it—past, present, and future. Hopefully, it will bring to readers a small portion of the joy and awe that walking the Cape brought to me.

Coast Guard Beach to Race Point
The Outer Beach

The idea of walking around Cape Cod began when my friend Gordon Wright suggested that we do "Thoreau's walk" from Coast Guard Beach in Eastham to Race Point in Provincetown. I quickly agreed. Intent on doing the walk in one day, we set off in the wee hours of one morning soon thereafter, accompanied by my daughter, Priscilla. Not having read Thoreau for 40 years, neither of us realized that Thoreau took three days for his trek and spent as much time off the beach as on it.

The cool night sky sparkled on this first walk and the ocean reflected the ambiance as we headed in the direction of the clearly visible north star and made our first tracks in the sand. Tide was low, hence initially we walked the waterline—seeking what Henry Beston labeled the "good footing along the water's edge." The beginning miles passed easily and pleasantly as we walked together, conversing easily. The roar of the surf, which can sound like a busy airport when the sea is up, was muted, not making me wonder (as I sometimes do) just how many drops of water must smash against the beach and how many grains of sand must grate against one another to create that deafening roar.

As the sun peeked above the horizon, there was a fog bank sitting offshore, blocking the sun's rise for a few minutes. Before the sun broke free, the ocean was black and cold—like winter— even more foreboding than it had been at night.

At low tide the bottom of a sandy beach is often flat, firm, and the site of the most interesting flora and fauna, the preferred place to walk even if one gets wet feet.

The area between high and low tide slopes and is normally looser, a far less comfortable place to journey, particularly on the open-ocean beaches where more active wave action results in a larger, coarser grain of sand. Although the terrain at water's edge does provide better footing for a brief period, prolonged exposure, it reminds me of one of the rituals of my New Hampshire boyhood, luring the younger lads out on a sidehill badger hunt. Sidehill badgers, as everyone knows, have shorter legs on one side than the other. Too much time on a sloped beach can become extremely uncomfortable.

The top of a beach is also flat, but often has even looser composition, thereby providing the worst footing of all.

As night gave way to day and both tide and temperature moved upward, our nighttime daydream was becoming a daytime nightmare. We trekked on, by now sometimes together, sometimes far apart. The two walking choices that remained—the skeletal contortions at water's edge, or the difficult footing at the top—both produced increasing discomfort. I soon realized that the most comfortable footing of all was right in Priscilla's footsteps—which somewhat limited the view.

By the time we stopped for a late morning lunch at Truro's Ballston Beach, carefully calculated to be a little over halfway, it had become obvious that completion of this walk was unlikely to leave us with a "warm and fuzzy" feeling.

By then the beach had turned into what Gordon described as a "seemingly endless treadmill—bluff on the left over which you could rarely see, narrow beach straight ahead, and unvarying ocean on the right." Whether viewed from Wellfleet or Truro, the view had come to seem more and more like the same picture showing up on every page of the same magazine.

We began to take breaks more frequently. Conversation was more strained. During one break when Gordon finished eating

Coast Guard Beach to Race Point

an apple and prepared to throw the core on the beach, Priscilla resisted, insisting that "we take out what we brought in." When I opined that the core was biodegradable and that "a seagull will probably eat it anyway," she responded that "we should not let the seagulls get off their normal diet." Come to think of it, the reason we didn't see many seagulls on this stage of the walk anyway, was probably because there were not enough humans around leaving apple cores. In one of the more pleasant moments of the walk's last few miles, Gordon and I interred the apple core while Priscilla was off "walking" in the dunes.

Towards the end, as we rounded each curve, we saw something that we thought might be Race Point. But, first we again encountered mankind in the form of a "dune taxi" out showing tourists the beach. One Patriot's Day after viewing both the reenactment of the battle of Concord and the Boston Marathon, I wondered just what it would have been like if the redcoats on their march out of Boston had come face-to-face with the marathoners running in. This day I wondered what Thoreau would have written about the dune taxi.

We resisted the temptation to pay our way to the finish, took our shoes off for some wading, and staggered into Race Point exactly on schedule at 5 p.m., sorer but wiser.

Thoreau's beach is among the most difficult beach walking on Cape Cod. Thoreau was no fool, terming the beach "heavy" and reporting that the "soft sand was no place to be in a hurry, for one mile there was as good as two elsewhere."

John Hay describes it this way in his book, The Great Beach:

> I plodded on noticing very little after a while, my attention blunted, reduced to seeing that one foot got in front of the other. The more level upper parts of the beach provided fairly good walking, but the sand was soft, and to relieve my aching muscles I would then angle down to the water's edge where it was firmer, and there I was obliged to walk with one

leg below the other because of the inclination of the beach. So I would return to the upper beach and push ahead.

Why, therefore, is the beach of Thoreau's and Hay's walks, and of Beston's Outermost House so famous?

Perhaps it is because one can walk over 25 miles without getting wet feet. Thoreau was aware of this, and it was far more important to him—carrying his earthly belongings—than it later was to me as I walked around the Cape, always prepared to swim. Maybe it is because in Thoreau's day this was where the action and drama were—ship wrecks, lighthouses, mooncussers, and rescue stations. Or, perhaps it is because this is the beach where land, sea and man come into the most continuous confrontation and the ocean often wins. Man has always liked an underdog.

We later calculated that we took sixty-six thousand steps in the sand that day in what Gordon labeled the "Seashore Struggle."

WALKING THE OUTER BEACH IN THE CAPE COD NATIONAL SEASHORE

My favorite walking area in the National Seashore is between Ballston Beach and Longnook Road in Truro.

The entire coastline is a straight scarp (eroding sandy cliff) well over 100 feet high in places, part of a continuous fifteen mile stretch of scarp. At the bottom of the cliff is a flat expanse of heavy loose sand, wider in the summer than in the winter. The walking is not easy between Longnook and Ballston or elsewhere on the outer beach; the sand on top is very loose and the slope to the water below is too steep to be comfortable.

There is a small section of the beach between Longnook and Ballston frequented in the summer by nudists. Those offended by such activity should avoid this walk.

The trail from the town parking lot to the beach at Longnook goes diagonally down the scarp. Walking down (or up) can be adventuresome in the off-season. At the other end, Ballston Beach, which was once the home of a life-saving station and then a Coast Guard station (that presently serves as a youth hostel), has more recently been the site of several storms. The storms

Coast Guard Beach to Race Point

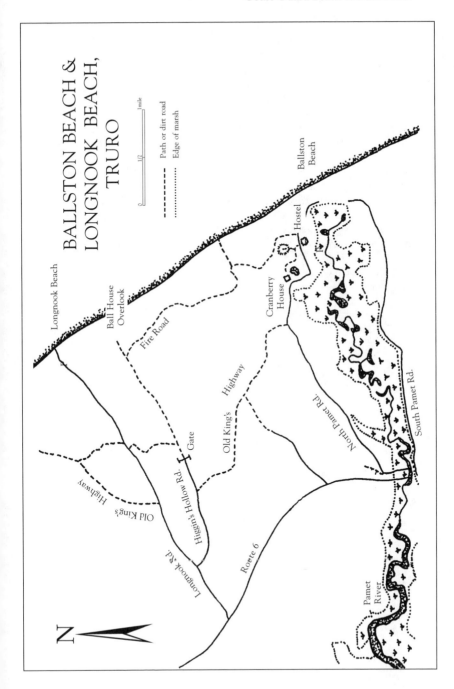

have washed over the dunes, threatening to link the ocean up with the Pamet River and turn those portions of Cape Cod to the north into an island. Sand washed in by these storms has buried the loop connecting North Pamet and South Pamet Roads, turning both into dead-end roads.

There are two routes slightly inland that a walker can take to avoid retracing steps along the beach. A fire road beginning just east of the hostel provides wonderful vistas of the Truro hills and the ocean. A steep hill to the left of this road (across North Pamet Road from the youth hostel) represents one of the best places on the entire Cape Cod perimeter to view the ocean. The hill also provides a view of the cranberry pond and house to the west, accessible by a separate trail maintained by the National Seashore. The hill is also topped by an unusually large expanse of bearberry.

The other alternative return route is a section of the Old Kings Highway.

The double valley of Higgin's Hollow and Longnook is unusual and attractive for walking in two respects. The hills are more reminiscent of northern New England than any other place on Cape Cod. And, creation of the National Seashore has preserved the local landscape as it was many years ago. Contrasted with most of the rest of Cape Cod, the area imparts a sense of long ago.

Provincetown
Zoning

In Provincetown, as in nearby Wellfleet, Cape Cod's history is still visible from the shoreline. The best way to get a panoramic perspective of the town is to walk on the flats at low tide. Even though Provincetown is a "deep" water port by Cape standards, much of its waterfront is a massive sand flat at low tide. Many of the buildings crowded along the shore are built right out over the water on wooden piers or posts. I have often wondered whether any of the people who have water under their living room at high tide have cut a hole in the floor to fish. No Levittown here, each structure is different from the next. Every building is so dissimilar from the next—some are different from any building anywhere—that in some ways they are all the same. Nothing is remotely modern about any of the structures, all display the reality that they have been there for many years. Piers are everywhere, a reminder of the days when the population made its living in the backyard—Cape Cod Bay.

The largest pier, town-owned MacMillan Pier, still serves as the terminal for ferries to the mainland, whale-watching boats, and other large ships. It needs some repair, several million dollars worth. Provincetown is a fiercely independent place, but the town recently passed a harbor management plan as the necessary first step to get state and federal aid to rebuild the pier.

Indeed, most of the piers have seen better days, now serving as reminders that it was once easier to make a living from the sea. There was a time when there were fish drying on every open space in Provincetown. In neighboring Truro, the remains of a platform are visible 100 yards or so out to sea off the town landing at Cold Storage Beach. This platform served as the unloading station for a cable apparatus to a former shore-front fish plant, so fishing boats could be unloaded at sea. In Provincetown, the large former icehouse, like so many other large structures on Cape Cod, is now a residential condo project.

Cape Codders have always made their living from the sea. It was never easy. The Pilgrims were perhaps the wisest. They got off the Mayflower, looked around, then got back on and proceeded across to Plymouth. They perceived Plymouth to have a better balance between the resources they could put to work for them and the natural dangers to which they would inevitably be exposed.

Despite this, Cape Cod was, of course, ultimately settled by European peoples. From the start, the population supported itself with what the sea provided by whaling, fishing, salt-making, mooncussing, rum-running, romancing and entertaining tourists, and finally by selling the land itself to wealthier people from off-Cape. Until this century, economic activity on Cape Cod was a never ending struggle between making a living and the very substantial risks of doing so, including death. Generations of men were lost at sea by being in the wrong waters during the wrong storm. Now, for the most part, storm damage is limited to losing a piece of the bluff—and whatever happened to have been built on it.

When necessary, our predecessors restructured the shoreline to make a living. They dredged harbors out of marshes, built dikes to change water flows, and constructed many varieties of buildings and piers right on the water's edge whenever and wherever there was money to be made. Imagine the uproar today if someone proposed to build salt-drying vats right on the beaches and marshes!

Until World War II, no zoning was necessary on Cape Cod. There were not enough people to get in each other's way, and, until Rachel Carson's Silent Spring called attention to the destructive results of pesticides, few people thought of the environment as needing protection.

For the most part, our predecessors did not choose to live on the shores. They saw enough of the sea while they made a living and preferred to dwell in the comparative shelter and serenity that was available inland. Until the current century, waterfront land was not particularly valued on Cape Cod. We are just now losing the old-timers who remember when their ancestors "could have bought a large stretch of the town's waterfront for a pittance."

Provincetown, surrounded by water on all sides but sheltered by the curl of the Cape's tip, was the hub of considerable historic activity. Although the economic purposes for which many of the properties on Provincetown's shoreline were created have long since ended, physically the town remains pretty much as it was.

Provincetown and Wellfleet remain alert for new methods of milking the sea to make a living. Whale-watching is big business and on the flats at low tide local citizens work their aquaculture grants, "farming" the sea with their trucks at their side. Even so, winter unemployment reaches 50 percent when the supply of tourists to sell T-shirts and souvenirs to diminishes.

One block back from the water, pedestrians still rule the crowded main streets of Provincetown, moving with a sense of invincibility that leaves motorists struggling in the belief that if there is a collision, somehow the pedestrian (or cyclist) will come out best.

Despite the extinction of the purpose for which many of the properties were created, there is much that is physically very appealing about Provincetown—a compact walking town surrounded by open space. But, if Provincetown had not been built when it was, none of it could be built now. We love our historic places, but now prohibit the construction of anything like them. A scene of Wellfleet appears on the cover of several Cape Cod

Commission publications, including one on "Designing the Future to Honor the Past," but, I wonder whether the setbacks, proximity to wetlands, density, or even the height of the church spire pictured would meet current codes. Why is it that we love history, but seem so blind to its lessons?

We live in an arrogant time, seemingly believing that after all those years of doing this and that, we have found perfection. We certainly make it difficult to change anything. But, most of the development of the past 20 years is among the Cape's ugliest.

Naturalist Robert Finch recently wrote in The Cape Codder:

> The sense of these former lives, and of former landscapes, is one of the things that has always made the Cape landscape seem to me one of the richest I know—perhaps because it suggests that we can live long and intensely in a landscape and not irreparably damage it.

Is modern grid zoning, one home per acre, an abomination? Not in need, purpose or intent—Cape Cod is getting more crowded—but perhaps so in the drab uniformity and comparative sterility that it produces. Whatever amount of space we plan to devote to such purposes as housing, retail, and making a living, we might be better off, like Provincetown, to cluster it in the old style. From my viewpoint, history and the Cape Cod National Seashore have treated Provincetown well, creating an intensely developed walking village surrounded by open space. Does anyone really believe that the charm, appearance, or utility of Provincetown would be improved by rebuilding it to meet today's zoning requirements and thereby sprawling it out over much of the outer Cape?

Today's zoning laws are as ineffective to the needs of Cape Cod, as a lack of zoning once was to it. It is time to begin the difficult task of changing them.

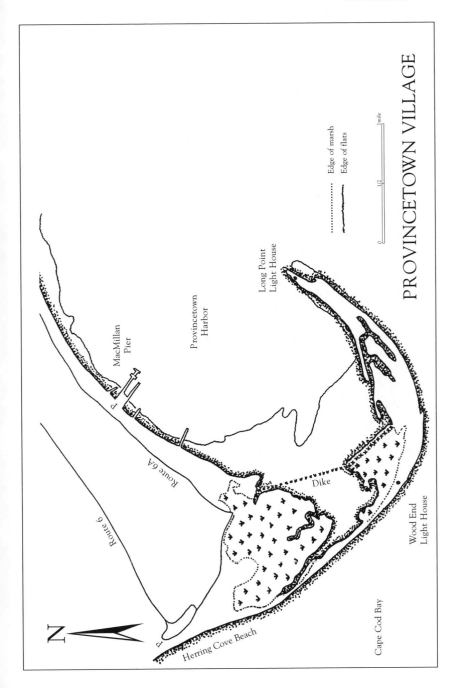

WALKING IN PROVINCETOWN

The shoreline of the Town of Provincetown offers a variety of attractive walks.

The town itself is best viewed from the beach and flats at low tide. Almost all of the waterfront of the town is privately owned, but it can be walked without difficulty at low tide—but ONLY AT LOW TIDE, providing an excellent view of the scenic and historic waterfront. Homes and businesses alike are built onto piers, right out over the water's edge.

The stone dike from the West End Circle to Wood End Lighthouse is also picturesque and historic. There is an excellent view of the harbor on one side and a salt marsh on the other. The dike represents an excellent example of the vulnerability of formidable objects to the power of the sea. Once topped by a flat stone surface, the dike now contains sections which are sagging and irregular. Portions can be under water in high tide during a storm.

There are two scenic lighthouses at Wood End and Long Point. The beach on the outer side represents the end point for the eroding sand being transported around and down the curving shore from Race Point. Most of the boat traffic in and out of Provincetown Harbor and much of the rest of Cape Cod Bay moves along this shore. Whales, seals, and many birds may be seen at appropriate times of the year. Although inland some sections of the windblown dunes and marsh are restricted due to bird nesting and replanting, in places there are paths along the dunes and the edge of the marsh that provide alternatives to the beach.

Moving south from Herring Cove Beach, there are informal sections for various sunbathers, including a women's section, a men's section, and a nudist section. The beach between Herring Cove and Wood End Light should not be walked by anyone uncomfortable passing through such areas. For those not uncomfortable, a facet of the character of the outer Cape is on view, including on occasion, the ongoing struggle between rule-enforcing National Seashore Rangers and the independent-minded local populace.

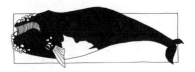

MacMillan Pier, Provincetown
Northern Right Whales

In Moby Dick, Herman Melville described the right whale as an "inferior creature—a lump of foul lard." Wrote Melville:

> If this whale be a king, he is a very sulky looking fellow to grace a diadem. Look at that hanging lower lip! What a huge sulk and pout is there! A sulk and pout, by carpenter's measurement, about twenty feet long and five feet deep, a sulk and pout which will yield you some 500 gallons of oil and more.

There are 11 species of great whales. Ten of the species, including the northern right whale, are baleen feeders. They eat by opening their mouths and trapping plankton on the hairy fringe of the strips of baleen, which serve as sieve-like strainers. The mouth of a right whale, which contains up to 260 baleen plates—each up to one foot by nine feet in size—is the most unusual and interesting part of the whale.

The eleventh species of whale, which has teeth, is the sperm whale.

The abundance of whales was once such that the Mayflower Pilgrims noted that "one could almost walk across" Cape Cod Bay on their backs.

Northern right whales grow up to 56 feet in length and 55

tons in average weight. In colonial days, they had the misfortune of being the wrong whales, the easy ones to catch, an ease they received their name from. They represent a combination of the ugly duckling, the tortoise who raced with the hare, and the village idiot.

Right whales look strange. Their barrel-shaped bodies have been labeled "a gross assemblage" by the late Jacques Cousteau. Their heads are disfigured by callosities, large discolored scab-like encrustations several inches thick that serve as a home for whale lice, barnacles, and other freeloading creatures. Their faces seem upside down, with the mouth where the forehead should be and eyes where the cheeks should be.

Right whales move slowly. Their fleeing speed, seven to eleven miles-an-hour, is slower than that of any other species of whale, some of which can travel as fast as 40 miles-an-hour. Right whales have been known to be attacked and killed by pods of hungry "killer whales" (orcas), which are approximately half their size.

The brain of a right whale weighs approximately seven pounds, less than that of any other whale species. The size of their brain in relationship to body weight is also comparatively small. What brain they have, they do not seem to use as capably as other whale species for their own good.

Two parts of a right whale's body were once considered worth risking the lives of boat-fulls of men: whale oil, used as a fuel, and baleen or whalebone, used in (among other things) ladies hooped dresses. As Melville commented, "It was in Queen Anne's time that the bone was in its glory, the farthingale being then all in fashion. And those grand dames moved about gaily, though in the jaws of the whale, as you might say."

Right whales have always been an integral part of the Cape Cod environment.

However, the roles of right whales in the economy and the environment have reversed themselves. Once common, they are now rare, an endangered species, the poster child of conservation license plates and, along with lighthouses, the two most popular symbols of Cape Cod.

Once man hunted and brutally killed right whales, now man protects them and sells tickets on boats to observe them, although whale-watching boats are presently prohibited from getting closer than 500 feet of a right whale. In this sense, too, right whales are symbolic of a bigger picture on Cape Cod. To make a living, man once ravaged the environment. Now, for the same purpose, man protects the natural environment.

Melville described the killing of a sperm whale in Moby Dick:

> The red tide poured from all sides of the monster like brooks down a hill. His tormented body rolled not in brine, but in blood, which bubbled and seethed for furlongs in their wake. The slanting sun playing upon this crimson pond in the sea, sent back its reflection into every face, so that all glowed to each other like red men. And all the while, jet after jet of white smoke was agonizingly shot from the spiracle of the whale, and vehement puff after puff from the mouth of the excited headsman; as at every dart, hauling in upon the crooked lance (by the line attached to it), Stubb straightened it again and again, by a few rapid blows against the gunwale, then again and again sent it into the whale.

As recently as 1966, the peak recorded whaling year, man killed 66,090 whales of all varieties. Ironically, some of the same techniques once used to kill whales are now employed to preserve them. In one instance, the two activities have been undertaken by succeeding generations of the same families. For both purposes, lines pulling buoys and sometimes boats are attached to exhaust the whales. When the whale becomes sufficiently tired, its pursuers approach in a small boat, originally to harpoon the whale to death, but now to disentangle the lines thought to be threatening a whale's life. The fisherman father of Dr. Charles "Stormy" Mayo, co-founder of the Center for Coastal Studies in Provincetown, briefly killed whales in order to make a living. Now, Mayo is the leader of the Center's rapid response disentanglement team.

Less than 350 northern right whales remain worldwide, each recognized by its callosities and known by name. The number has not rebounded in recent years despite designation and protection as an endangered species. The whales continue to die as fast as they breed. The deaths are thought to be primarily due to collisions with large ships in the Great South Channel shipping lanes southeast of Cape Cod and elsewhere, although entanglements remain a concern. The reasons behind the lack of regeneration are unclear. Some observers believe that all existing northern right whales are now descendant from only three females, weakening the gene pool. Others worry about the food supply. From December to April, Cape Cod Bay is one of the major feeding grounds, as there is usually an abundant supply of the zooplankton which the whales eat.

Provincetown and Truro, although eventually eclipsed by Nantucket and New Bedford, were the pioneers of early whaling. Indeed, Cape Codders taught whaling to the other two ports. In the eighteenth century, the male population of Provincetown was substantially diminished when all whale boats were at sea.

I interrupted my walk to take a whale-watching cruise to Stellwagen Bank from MacMillan Pier—then and now the hub of Provincetown's whaling activities—and to talk with the naturalists at the Center for Coastal Studies about their research on the continued presence in Cape Cod Bay of right whales. In the summer, whale-watching brings 4000 people a day into Provincetown, clearly a very significant continuing factor in the town's economy, as well as that of the rest of the Cape.

Modern technology threatens whales in numerous ways, many not fully understood. For example, the U. S. Navy is considering utilizing a low frequency sonar device to detect submarines for anti-terrorism purposes. The device, dropped 100 feet below the ocean surface, reaches 100 miles with a bang—not a ping—of 90 seconds duration, operating on the same frequency as right whales hear. Although the impact is not known, the sound is sufficient to kill people in a submarine right below it. A deaf whale is a dead whale. In a sign of how much attitudes have changed, however, the U.S. Navy recently offered to assign one of its whale research vessels for a year of environmental impact studies before the device is put in use.

Provincetown's Center for Coastal Studies continues to study the right whale and to try to protect them through procedures such as the rapid disentanglement team. The Center, recognizing the need to share the ocean with a wide variety of other legitimate users, attempts to develop and utilize balanced procedures to protect the whale without disrupting other necessary activities, such as a current plan to monitor the location of whales in the Great South Channel during the migratory season so that ships may be notified of their location.

Every reaction, however, begets a counter-reaction. Just as society once wantonly killed too many whales, one extremist organization, Green World, has sued the Commonwealth of Massachusetts, the U. S. Coast Guard, and the National Division of Marine Fisheries, utilizing the Endangered Species Protection Act and other laws in an attempt to force them to restrict the use of lines and nets by lobstermen and fishermen in bays frequented by right whales. It has also initiated litigation against the major participants in whale-watching to keep them away from right whales entirely. These suits have the potential to inflict considerable economic damage on the affected industries, although there is little evidence that either lobstering or whale-watching has harmed such creatures. Knowledgeable observers in fact believe that whale-watching has helped by substantially increasing the number of individuals interested in protecting whales.

I recently sat in while a group of scientists discussed whether the whale-watching industry should be regulated. There was universal agreement that no problem actually exists, because, in the words of one naturalist, "whales are extremely capable navigators who take necessary steps to avoid a collision." (In this instance, the reference is to humpback whales and the other species viewed primarily by the boats, not to right whales.) Nonetheless, they believe that the industry should police itself to avoid a major public relations problem.

Lobstering is presently prohibited for much of the spring in areas such as Cape Cod Bay when right whales might be present. Unfortunately when the whales—which are now closely watched—left the bay earlier in the spring of 1997 than ever before, regulators were slow to permit earlier lobstering than previously planned.

In the continuing see-saw of relationships between multiple interest groups, the greater the temporary success of the overprotection movement, the more inevitable and severe will be the next round of threats to the "wrong" whale. Indeed, the U. S. Supreme Court recently ruled that parties damaged by the Endangered Species Act have legal standing to complain if they are unreasonably deprived of the value of their property, thus opening the way for the next counter attack by groups such as lobstermen.

A recent column in Forbes Magazine by Russell Seitz, an affiliate of the John M. Olin Institute for Studies at Harvard University, asserted that the "burgeoning" population of whales is among other results, causing numerous "collisions" with transatlantic yacht racers and "muscling aside" cod and other valuable food species in the North Sea.

The life of a northern right whale is not an easy one.

Truro

*Disagreement &
Due Process*

Day's Cottages on the bay side in North Truro are perhaps Cape Cod's most beloved and sought after summer lodgings. Resembling a row of Army barracks, the 24 small and identical, rectangular cottages—evenly spaced on the water's edge—are each named after a flower. When the wind is right, a lone sea gull perches as sentinel on the peak of each. It is not always easy to get a reservation; many have been passed down through families and friends.

Walking on the water side, it is easy to understand the appeal of the cottages, they are the next best thing to a houseboat or camping on the beach. Surfcasters and swimmers need walk only a few steps from the "back" door to enjoy themselves. On a stormy night the occupant undoubtedly gets the sensation that the waves will come right through.

Like Provincetown and so many other charming places on Cape Cod, literally nothing about the cottages would qualify under current building codes. They are too small, too close together, too close to the water, and sit right behind a series of groins atop a seawall which has probably preserved their existence. It would be next to impossible to get permission to build Day's cottages now.

The United States is a highly independent nation in which people have strong and differing opinions on many subjects. Everyone has a cause, but the process of working things out becomes continually more complex. The shoreline of Cape Cod is no exception. Everyone wants a piece, including boaters, swimmers, fishermen and shellfishermen, naturalists, homeowners, walkers, and commercial property operators. As David Dutra, Truro fisherman and sea scallop farmer put it at a meeting explaining his aquaculture grant, "The problem is that we have a lot of user groups."

Truro, the most thinly populated town on Cape Cod, would seem an unlikely place for conflict. Its Cape Cod Bay shoreline—a continuous sweep of sandy beach—is divided into two sections. Most of North Truro contains other motels and cottages, like Day's, built years ago right on the beach. No one seems sorry that the shoreline exists the way it is. The southern stretch, also beautiful sandy beach, is unbroken by revetments, docks, or any other disruptions, but is the site of a slowly growing number of large second homes, many sitting far above the water on top of bluffs.

The Truro shoreline is characterized, however, by constant disagreement concerning proposed change in any section.

The shoreline between Provincetown and Truro was once divided by an inlet into Pilgrim Lake. When a railroad was built in 1873, a dike built under the tracks stopped the flow of water into the lake. Since the water current was eliminated, the adjacent bay shoreline in Truro has been building up, there now being several hundred more feet of sand than before the dike.

Almost immediately after crossing into Truro, one passes the Sea Gull Motel. Its owners were participants in 1994 and 1995 disagreements over use of the shore. In 1994, they sued the State Division of Marine Fisheries in State Supreme Court and won, reversing the town's decision to grant aquaculture (shellfish farming) permits on the tidal flats in front of the motel to a local shell fisherman. The issue was whether aquaculture is permissible "fishing" under the Colonial Ordinances of 1654 which still control the use of such land. The court's answer: "farming is not fishing."

In the second dispute, the motel owners, without a permit,

built a ramp over the accreted sand to a roofed sun-deck right on the shore. They cited as justification their legal obligation to provide equal access to the handicapped, in the words of a local newspaper, those "who are on medications that prevent them from sitting in the sun."

The sun-deck, a somewhat flimsy structure moved on and off the beach each summer, does no visible harm. This leads me to wonder why our society finds it necessary to prohibit such actions in the first place, but sponsors of legislation to protect the disabled certainly never had this structure in mind. Initially, the Truro Conservation Commission denied permission for the structure citing other conflicting laws, but the State Division of Environmental Protection overruled. The case illustrates two principles of polarized conflict. First, if you look hard enough you can probably find a plausible legislative justification for any purpose—whether intended or not. Second, if the ruling of one regulatory authority is unfavorable, try another. There are plenty to go around.

Continuing southeast, a walker comes to Beach Point, home of Day's Cottages and its siblings. Until early 1997 skirmishes over the beach front here were relatively minor, dealing with issues such as whether or not to replace sand lost to the ocean from beneath seawalls. But, a January 1997 storm destroyed or critically damaged several bulkheads, thereby placing a large cloud over the future of many of the rental accommodations—Truro's "cash cow." In an awesome display of the power of the sea, segments of the walls looked as if a group of bulldozers had driven over or through them. The storm created considerable potential for conflict between naturalists and property owners anxious to take action to protect their source of income.

In 1996 I had walked much of the Truro shore—including portions of Beach Point—with Graham Giese a Truro resident and the Cape's preeminent geologist and expert on the natural movement of sand. Geologists can sometimes come across as overly academic and totally lacking in any pragmatic sense—as if they considered Cape Cod's shore to be one big laboratory experiment which should be allowed to play out for ten thousand years for their benefit. But, every walker should be so lucky to have such a knowledgeable companion.

Giese sees things that the amateur eye could gaze at forever and never see—like the windblown sand with beach grass growing atop that is piled up one story high on the undeveloped lots irregularly located amongst the cottages and motels of Beach Point. If only there were a more even mix between developed and undeveloped lots, he said, the unoccupied ones might catch enough sand to offset that being lost in front of the seawalls. Seems obvious once you think about it.

Truro's initial response to the problems created by the storm was positive and constructive. It created a new committee to review the situation, the members of which included both Giese and other conservationists and several of the impacted property owners. Hopefully the committee will find common ground.

As the walker moves south past the town landing at Cold Storage Beach towards the large houses on the bluffs of the Corn Hill section, the shoreline action again changes. Severe erosion in December of 1995 and January of 1997 consumed large chunks of the bluffs, seriously threatening several properties and leading to the condemnation of one left perched dangerously close to the edge. In the spring of 1996 following the storms of the prior winter, the face of the bluff beneath the condemned house contained three or four large "cracks" where the sand (and vegetation thereon) had split. By fall these sections had all come tumbling down, leaving a shear wall of fresh sand.

As they inevitably do anywhere on Cape Cod, the property owners sought methods to preserve their houses. Two, including the owner of the condemned property, initially proposed to build revetments. The Truro Conservation Committee, concerned about the long term impact on the beach of such piecemeal solutions, declined. In December of 1996, the condemned house—large enough to require the services of the same company that had moved Highland Lighthouse earlier the same year—was moved back. The second property has been sold and may also be moved back. Meanwhile, one new house is under construction and one additional lot on the same bluff has been advertised for sale in local papers, at a hefty price, with aerial photographs clearly showing the sections of eroded bluff in the area. Giese indicated that the lot extends far enough back so that if a house is positioned right, an owner could live there happily for many years.

Throughout this stretch, eroding sections alternate with places where the shore is building up. The erosion seems random. Giese attributes this to the irregular movement of offshore sandbars.

Approaching the mouth of the Pamet River, controversy moves about two miles out to sea. It is here that Judy and David Dutra, a Truro nurse and fisherman—two very ordinary Cape Cod people in the best sense of these words—are utilizing a $111,430 grant from the National Marine Fisheries Division to develop the nation's first sea scallop aquaculture program on ten acres of Cape Cod Bay. Ironically, even though the program is government funded, it took the Dutras three years to get the necessary town, state and federal government approvals. When it comes to aquaculture, the United States is the equivalent of a third-world nation. The problems of the Dutras' grant illustrate why.

No one outright opposed their proposal, but it did catch the attention of several other groups with differing interests. Right whale preservationists, concerned about the potential for whale entanglements, required total redesign of the equipment from that utilized in similar programs in Japan and elsewhere, necessitating that the Dutras employ trial and error experimentation in design and equipment every step of the way. Some Wellfleet scallop fishermen complained about the loss of the area to open fishing. This problem was aggravated when 400 acres were mistakenly marked off, not the 10 approved. In the heat of the dispute, some of the scallop cages were "mysteriously" cut adrift.

Again heading south, one final scene of recent disagreement awaits the walker of Truro's shores—whether or not to dredge the Pamet River and enlarge the dock in the River's Harbor to facilitate handicapped access. Probably neither project would have even been considered were it not for the availability of state funding.

No one outright opposed dredging the river undertaken in 1995 to facilitate passage by boats at all tides, at a total cost of $550,000. Skeptics just thought it wasteful, there being only 120 boats in the harbor (almost all of which are pleasure boats), and with considerable doubt about how long the channel would stay dredged. Since the dredging, considerable sand has in fact re-accumulated in the channel, leading to further debate concerning what to do next.

The pier was another matter. Local opponents held its construction up in court for a considerable period of time for similar reasons, and it is yet to be built. The state, wielding the club mandated by its funding, insists on a large pier with handicapped access. In the spring of 1997 40 neighbors petitioned the town to ask the state to cancel plans for the pier, because the pier would be so big that winter ice will rip it apart, and the town will be forced to pick up the cost of repairs.

I left Truro thinking that we probably could all get along better if we just worked a little harder at listening to each other, but far less certain that this would in fact happen. We live in a society where disagreeing "sells newspapers," and getting along does not.

WALKING CAPE COD BAY IN TRURO

Most of the Cape Cod Bay shore of Truro is privately owned, hence, walking is subject to the Colonial Ordinances and the good will of the property owners. Most of the shoreline is accessible at low tide. The northern end represents an interesting example of traditional Cape Cod motels and cottages. The southern portion has upscale private homes.

Truro

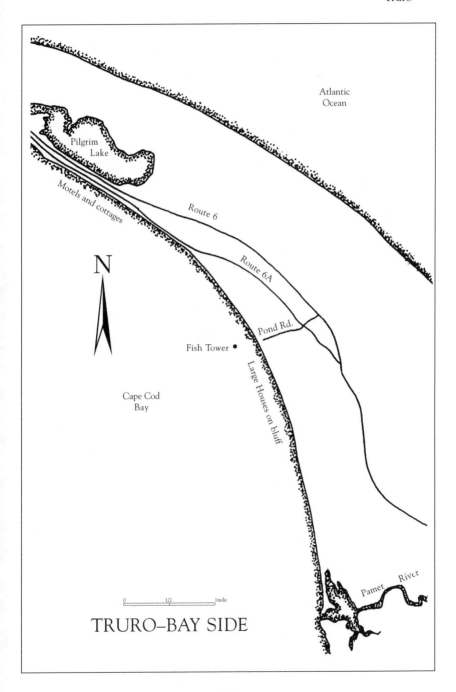

Great Island, Wellfleet
History on the Shores of Cape Cod Bay

The depression in the ground, a small indentation partly overgrown by vegetation, does not seem different from hundreds of others. But, this one is different, it constitutes the remnants of a 1970 archeological excavation to explore the site of the Great Island (Smith's) Tavern.

Cape Cod has always been a maritime society. At various times in history its shores have been covered with structures to support fishing, whaling, salt making, shipping of various types, and—some would say—smuggling. One writer has called Provincetown the richest town in America during the whaling era, and another stated that during the packet boat and clipper ship era (1815-1860) there was no more prosperous part of the entire national coast than Cape Cod. The Cape's shoreline almost always has been a beehive of activity. As Mary Rogers Bangs states in Old Cape Cod, the sea "was a friend to be loved, an enemy to be fought, a giver of food, and a solemn harvester that brought dead men to the door."

But inasmuch as tides sweep the shoreline clean twice each day and erosion constantly changes it, the debris of abandoned activities disappears faster on the ocean's edge than anywhere inland. As a result, although once many structures such as thousands of salt-making vats lined the coast, searching for their remnants is akin to looking for the proverbial needle in a haystack.

Cape Codders have a tendency to describe history as they would like it to have been—some prefer swashbucklers and some prefer Puritans. It should thus come as no surprise that at least three versions circulate concerning what went on at Smith's Tavern on Great Island.

A National Seashore sign at the parking lot for Great Island states:

> In the late 1600's, whales were so plentiful here that they regularly appeared in Wellfleet Harbor. Lookouts posted on high ground alerted the whalers who pursued the mammals in small boats.
>
> Local tradition holds that whalers often retreated to a tavern on the secluded bluffs of Great Island. Here they could recount their adventures, fortify themselves with toddy, and consort with sympathetic ladies while whale-watchers kept vigil nearby.
>
> During the summer of 1970, archeologists from Plimouth Plantation and the National Park Service located and excavated the foundation of a building on Great Island that may have been such a tavern. More than 24,000 artifacts were uncovered. Evidence indicates that the building was used between 1690 and 1740.
>
> The Great Island Trail which begins here leads to the tavern site. There are no visible remains.

No one disputes the fact that Great Island was one of the primary centers of whaling activity at the time.

However, in A History of Billingsgate written in 1991, Durand Echeverria has a more benign viewpoint than that of the National Seashore sign concerning the use of Smith's Tavern, concluding that "there was no dwelling on Great Island previous to March 1749," that "the main structure... was intended not for use as a public tavern but as housing" and that "the conversion (and perhaps enlargement) of the original dwelling into a tavern, which does indeed seem probable from archaeological and oral tradition, must tentatively be dated no earlier than the late 1750s."

Kenneth Kinkor of the Expedition Whydah Sea Lab & Learning Center in Provincetown, believes that Great Island and Provincetown served as transfer places for smugglers in the 17th and 18th centuries, and cites "the unusual amount of storage space" in the tavern as possible evidence of its use for that purpose.

All three parties refer as support for their viewpoint to the dating of pipes made in England and uncovered when the site was excavated. The pipes were made early enough to fit any of the three theories, but could have been left there at a later date.

Although all of the Cape towns engaged to some varying degrees in many of the Cape's traditional maritime activities such as fishing and salt-making, history has come to link certain towns with special activities. The outer-beach towns from Chatham to Provincetown are most closely identified with shipwrecks, mooncussing, and rescue. Provincetown, Truro and Wellfleet are linked with fishing, whaling, and, some would say, smuggling—Kinkor says Provincetown has always been a law unto itself. As one turns the bend on Cape Cod Bay, the villages of Brewster, East Dennis, Yarmouthport, and Barnstable are more known for salt-making, packet boats, and clipper ships, the commerce of the nineteenth century. Brewster was known as the "ship captains" town, having provided as many as 50 ship captains at one time (more than from any other town in the entire country) from a total population of only slightly in excess of 1,000. Brewster boys went off to crew on the clipper ships at the age of about ten. As many as three quarters of the male inhabitants were employed at sea in the peak of the clipper ship era. Ironically, the most vivid living monuments to the Cape's maritime tradition, the homes (with their famous widow's walks) and the churches built by the ship captains (with their names at the end of their pews), are not on the coast at all, but scattered along the main thoroughfare, then known as the Old King's Highway, but now better known as Route 6A.

Brewster also offers one of the most substantial visual remains of the packet boat era, a large stone breakwater built on the sand flats. The breakwater was designed to shelter a hole dug on the leeward side, in which a boat could continue to float at low tide. Two large posts from the pier which once led to the breakwater also continue to appear and disappear in the shifting sands at water's edge. Many boards which might have been part of the wharf lie on the flats at low tides. I have asked many people what their historical origin was, but none are certain. Some cedar stumps are thought to have survived for as long a thousand years at water's edge, and I can think of little other purpose for these boards.

Why, one wonders, did Brewster, the most sand-locked of all Cape Cod towns, emerge as a leader when it came to the biggest and fastest boats put to sea manned with Cape Cod seamen? Was it perhaps because the able-bodied men from such places as Provincetown and Chatham, more conveniently located to maritime trades, took all the good local jobs, forcing the poor Brewster lads to go much further afield to make a living?

The plaques in East Dennis which commemorate the location of the Cape's only clipper-ship-building yard and the peninsula's first saltworks are less than a quarter mile apart as the crow flies, but considerably further if one has to walk from one side of Sesuit Creek to the other to view both. The Shiverick shipworks built six clipper ships in the 1850s on the edge of the creek's salt marsh, and somehow managed to launch them all on the flood tides of late fall or early spring.

Dennis had more saltworks than any other town. Saltworks were complex structures which covered much of the coast when the industry was at its peak. Windmills lifted the water from saltwater creeks into wooden pipes which carried it to a series of three wooden tubs where the water was allowed to evaporate off, leaving only the salt in the third vat. Whenever it rained, the population of the village would run to cover the vats, so as to not retard the process by allowing any more water to accumulate.

After several years of asking and looking, I found one remnant of this industry, the ends of a hollow wooden pipe approximately 12 inches in diameter which emerge from both sides of a creek in the salt marsh, about two feet above the low tide line. The pipes are thus submerged for more of the tide cycle than they are not—another very real example of the lasting power of certain types of wood when preserved primarily in peat and saltwater.

The final stop of my search for history on the shores of Cape Cod Bay was in Sandwich to view the remains of the Sandwich Glass Works, a large factory built at the water's edge in 1826. Part of a small furnace is all that remains of the actual plant. However, two large straight channels were dug at right angles through the salt marsh to facilitate navigation right up to the factory. They remain, now looking just like all of the other channels flowing through the marsh. If it were not for their straightness, they would probably be unidentifiable. Dredging is one of many tasks which our ancestors seemed to perform with both more skill and enthusiasm than we do now. No one seems to be exactly sure how they did it, one theory being that they dammed off an area and then utilized oxen to plow it out.

As one walks Cape Cod Bay it is striking how little remains from all of the activity which once took place on its shores. The shore has a remarkable ability to cleanse itself in a short period of time, with the result that little long-term damage occurred on the Cape coast itself. Ironically, far and away the most serious environmental damage man's activities inflicted here was the denuding of nearly all the inland portions of the peninsula, in order to provide the wood necessary to fire the furnaces and build the structures.

Stripped of trees, the topsoil blew away, making inland Cape Cod into a far more barren land than it was when the Pilgrims landed.

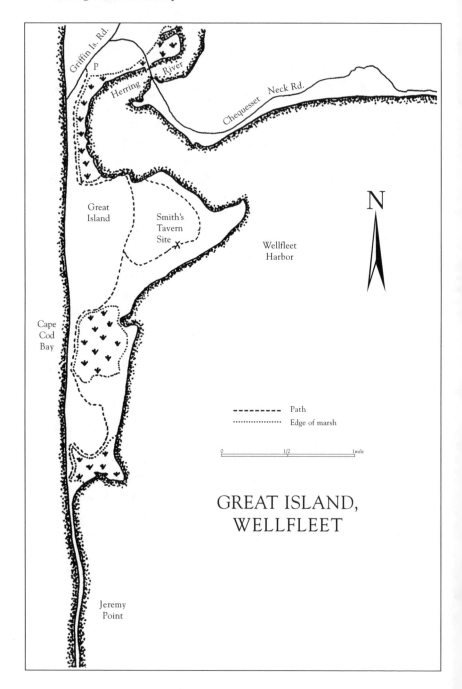

Great Island, Wellfleet

WALKING GREAT ISLAND IN WELLFLEET

Great Island in Wellfleet presents three options to the walker:

- The beach on the Cape Cod Bay side, created by the erosion process, is typical of Cape Cod: straight, sandy, and uninterrupted. The last mile, Jeremy Point, is a sand spit.

- On the opposite side there is a circuitous route along a marsh and the island with an excellent view of Wellfleet Harbor and the Town of Wellfleet itself across the water.

- A trail down the middle of the island winds its way through a pleasant section of woods. Uprooted trees in several areas show the power of the wind on the exposed shores of Cape Cod. A side trail leads to the site of historic Smith's Tavern.

Campground Beach, Eastham
Erosion

After a winter storm, Campground Beach in Eastham is the best place on Cape Cod to view the impact of erosion and man's efforts to resist it. The edge of the parking lot is jagged, scarred by the large chunks which were torn away by a storm in December of 1995. A similar storm the following winter spared this parking lot but took similar pieces out of the lot at nearby Silver Spring Beach. To the north of Campground, the shoreline has been protected for many years by a rock revetment (large rocks laid into the bank to diffuse the wave action). It extends approximately 30 feet into the bay beyond the unprotected parking lot. As a result, the beach is impassable to a walker at the high end of the tide cycle when water reaches the revetment. Aerial photographs of Eastham confirm that the beach is much thinner below the revetments.

To the south, the shoreline has been protected only by so-called "soft" solutions: sand bags, snow fences to diffuse water action and gather more sand, and plantings of beach grass and other vegetation to hold the sand. This shore is on a line with the unprotected parking lot. Large sections of the bluff caved in during the severe storms of the past two winters, carrying snow fences and stairways down to become twisted debris in the sand below. At the top, several cottages always stand ominously close to the edge after each storm, often with decks hanging over the edge, the outermost underpinnings suspended without support in mid air. Cottages that can be moved back soon are.

My friend Noel Beyle lives on the bluff in this section. His home is fifteen feet from the edge. There is no view. As the wind hits the bluff, it lifts sand upward in a classic display of the power of natural events, and then deposits it on the top as the updraft disappears. The resulting pile occupies almost all of the space between the house and the edge, rising above window level.

Noel loves the beach. Inasmuch as his home cannot be moved back, he is the only owner I know willing to leave his home "in the line of fire" to avoid building a revetment which he believes might "injure neighbors' property and scour out the beaches until there's little left to walk on." Instead, Noel believes that "soft solutions should be encouraged in the erosion fight" and spends considerable time and effort working on his own section of the bluff. He has not lost any land during the severe storms of the past two winters, but does not disagree vehemently when I imply that fortuitous, and temporary, placement of sandbars might be the reason.

Noel and I do not always agree when we discuss erosion, but I respect him. His bottom line is that, "we need more stewardship of land at the edge of the sea, something worth handing to our children—beaches, not rock walls; public assets that drive the economy, not private fiefdoms benefiting private ownership. Private rights have limits, and they should begin at the water's edge… mean high."

Eastham's Campground Beach represents a transition in another respect. To the north, the shoreline is most vulnerable when the wind blows from the southwest. The prevailing current then flows north along the shore carrying eroded sand in that direction along the revetment. To the south, a northwest wind travels further while crossing Cape Cod Bay with the result that the prevailing current and sand move southward. One of the determinants of how large (and strong) waves can get is the fetch—the distance which they can travel and continue to build.

Campground Beach, Eastham

Campground Beach is a dividing point in respect to the direction waves come from in traveling the furthest distance across the bay. As a result, sand being scoured out of Eastham's Cape Cod Bay shoreline is being carried to and deposited at both ends.

One of the reasons erosion has been worse in recent years to the south of Campground Beach than to the north is the disappearance of what once was Billingsgate Island in Wellfleet. The island, positioned to the south of Jeremy Point at the end of Great Island in Wellfleet, was once large enough to hold a lighthouse, an inn, and a small village. It washed away early this century, now being survived only by a pile of rocks at the base of the former lighthouse. Its destruction lengthened the fetch striking the Eastham shore from the northeast.

Much of Eastham's bay shore has eroded over the years into a straight sandy bluff several miles long. One longtime resident told me that 61 feet have been lost from his property in the past 50 years, approximately a third of the rate of loss in the more publicized erosion at Nauset Light on the other side of Eastham.

If the process of resisting erosion is considered a battle (revetment builders refer to "armoring the shore") here it is clearly "everyone for themselves." For much of the shoreline, everything but the kitchen sink has been thrown in to temper the sea, one piece of property at a time, in a highly uncoordinated fashion.

A large number of irregular revetments, wooden bulkheads, and snow fences accompanied by beach grass are intermixed with properties where nothing has been done. The two most unusual solutions are a pile of tires left on the beach to hold sand, and a Rube Goldberg-like turnstile fence, built entirely of homemade parts. In its entirety, this shore of Eastham, as a result of many years of changing policies, leaves one with the impression that the plan is to have no plan.

Erosion is a predictable process. Except for the unusual hurricane or nor-easter, erosion is worst during "spring tides" unusually high tides which occur twice a month and are worst during the winter. It is all so predictable, one could almost sell tickets to tourists to watch houses fall into the sea. We know when, and sometimes even at what time of day, it is most likely to happen.

What we don't know, however, is which way the wind will be blowing and pushing the surging power of the water. When it blows from the northwest or southwest during such tides, erosion takes big chunks out of the north and west facing shores of Cape Cod Bay, particularly in Eastham.

The Eastham shore is a living laboratory of what works against erosion, how it works, and what does not—the results being quite dramatic. Wherever there is a revetment, little land has been lost, and wherever there is no protection or soft solutions, large chunks of the bluff are gone after a major storm. The entire shore remains in parallel straight lines, but the portions protected by a revetment, regardless of the length, now protrude 20 to 30 feet to the west of the remainder of the shoreline.

In spite of their comparative effectiveness, however, revetments and other "hard" solutions are not universally beloved. Indeed, there is intense debate about their appropriateness and it has become increasingly difficult to get a permit to build one. Geologists and their adherents believe that it is inevitable that erosion will eliminate Cape Cod entirely in approximately 10,000 years, and that it is inappropriate and offensive to do anything to attempt to stop or slow the process.

But, "a man's home is his castle" and many want to do anything, including building revetments, to protect their castles—even if just for a lifetime and not 10,000 years.

The intensity of the debate on both sides of this issue was brought home to me one day recently when several long-time Cape Codders opined that "there never would have been a problem if we had sunk a few battleships" in the 1987 cut in the outer beach which caused severe erosion problems in Chatham. Later the same day, a more recently arrived retired couple from off-Cape expressed the belief that all those wealthy people who "foolishly build houses on the water's edge" are getting "what they deserve." When I countered that several of the homes lost in Chatham had been there for many years and that not all of the owners are necessarily wealthy, they switched to "well then they didn't pay much for the houses anyway." Both groups are well-educated and otherwise fully rational.

A third group—officials at the town, state, and federal level—get politically whipsawed over the "winner-take-all" struggle of these other groups, making it increasingly difficult to provide consistent leadership. For example, for a few years after the break in 1987, Chatham selectmen viewed what was going on as natural and evolutionary, and opposed revetments. In 1996 a new board played a leadership role in building one, ironically at town expense, on the property of several homeowners who once sought to build such protection at their own expense.

Unfortunately, in such a politically charged environment, it does seem to make a difference whether the next house falling into the sea belongs to an out-of-town owner who would not let the local folks walk on the beach, an elderly widow who has lived there for many years, absentee owners who rent their property, or a town official.

Although erosion on Cape Cod and efforts to combat it are probably not very important in the global scheme of things, they thus represent an example of the increasing difficulties in America of finding a moderate and rational solution to any problem. When two groups of extremists fight, no one is served. "Why can't we all just get along?"

One of the things I came to realize as I walked around Cape Cod was that although I am not a naturalist, a geologist, a lawyer, or a shore-front property owner, I may be the only one who has seen every revetment, seawall, snow fence, and eroding bluff on Cape Cod. What I viewed was quite different from much of what I had read. I have seen quotes from officials of some towns that bear so little accuracy to what I saw from the shoreline of their own towns that I wonder if some of them have ever been there.

To summarize my observations briefly:

- Despite regulatory discouragement, owners continue to build in inappropriate places. As noted, not too far to the north of Eastham, in Truro several expensive homes built in fairly recent history on large mounds of sand were surprised by erosion in the same winter storms of 1995-96 which damaged Campground Beach. Although I have considerable sympathy for the owners of a home built many years ago, and consider it a serious political mistake of conservationists to show so little concern for property owners, it is hard to have sympathy for owners who built or bought their properties long after they should have known better. A bill proposed in the 1997 session of the Massachusetts Legislature would require fuller disclosure concerning the history of erosion in the area of any property being sold on the waterfront.

- In terms of the life of man, if not the life of a geologist, some hard solutions do work. Nothing works forever. The wrong storm from the wrong direction at the wrong tide may well overcome anything that man can erect. But, seawalls, revetments, and other hard solutions have obviously protected neighborhoods in Cotuit, Falmouth, Bourne, and elsewhere for many years. Some are ugly, some add historic character to the area. In Hyannis, for example, one large house built in 1906 occupies a massive seawall seemingly constructed at the same time, the beach having eroded well back on both sides. On a bend in the Mitchell River near Stage Harbor in Chatham, a house cantilevers out over the seawall on which it is built. On a larger scale, such landmark properties as the Provincetown Inn and the Seacrest Hotel in Falmouth sit partly on shore-front mounds of concrete and stone. These moderate successes of hard solutions should be no surprise. Rocks do occur naturally in the environment. On Cape Cod wherever there are rocks, there is a point.

- On the outer perimeter of Cape Cod, soft solutions are not much more than wishful thinking, perhaps a rationalization with which to avoid a hard solution. When not needed, they do improve the cosmetic appearance. But when a real storm comes, as it did to Eastham in each of the past two winters, they might as well not be there. In The Sea Around Us, Rachel Carson tells of the sea once moving a concrete breakwater weighing 2,700 tons at a port in Scotland. Does anyone familiar with this power really believe that a snow fence will do much good anywhere other than in a sheltered cove?

• Every element of the Cape's shoreline is different: tides; prevailing wind direction at various times of the year; currents; how far a wave can travel and grow before it strikes land; and the geological composition of the shore all vary. For this reason, what is true about erosion in one area can be quite false in another. Too many statements made about one place are later assumed to be universal truths or misapplied to another place.

• Finally, whatever is done—hard, soft, or nothing—it should be consistent. If there is to be a revetment, there is a logical starting point and a logical ending point. Irregularity serves no one for very long. A betterment-like process should be required for logically defined areas if any action is to be taken on the shore front. Eastham's shore represents sad testimony to this reality as do those in Chatham and Dennis. In both towns long stretches of revetments are broken up by gaps which may ultimately threaten the life of the entire system.

If revetments do provide some protection, why are they so controversial? Opponents make three criticisms: they "scour" the beach at both ends; they "scour" the beach in front; and they "starve" the shore elsewhere by impeding the movement of sand. The common theme in all is that hard solutions are a threat to beaches—the pride of Cape Cod. The primacy of beaches in such criticisms conflicts somewhat with the legal reality in Massachusetts that the public has limited right to be on private beaches anyway. In Eastham and elsewhere I saw very little evidence of the first two criticisms.

As long as revetments are parallel to the shore, there is little or no scouring at the ends. As noted, revetments in Eastham and elsewhere do protrude further to sea than adjacent property, but the property fronts are remarkably parallel whether the revetments or unprotected portions of the shoreline are one mile or 100 feet long. On my walk, I only saw end scouring where a revetment angles into the adjacent shoreline.

There was little consistency to what I saw in front of revetments. In some places sand is building up and in others there is little visible change. In Eastham, there is less beach, about 30 feet less, because that is approximately how much land the revetment has preserved. Although I did not measure, which would be easy to do, I saw little visible evidence of revetment induced variation in the height of the beach . Some aerial photographs of Eastham, however, do show a lesser buildup of sandbars off shore in front of revetments. At low tide I occasionally had to choose between scaling along the side of a revetment at an awkward angle or sloshing knee deep in water below it, as I did at Eastham at high tide. But it was my impression that the depth was due to the reality that by sticking out further, the revetment reaches deeper water. The longer a revetment lasts, the deeper the water. This does reduce the width of beaches and eventually eliminate their use entirely at portions or all of the tidal cycle.

It seems an obvious truism that if revetments slow or deter erosion, they stop sand from going elsewhere. Is that not the

---

WALKING EASTHAM

Much of the shoreline of Eastham is privately owned. Hence, the walker again is subject to the Colonial Ordinances or the good will of the owner.

The area north of Campground Beach, protected by revetments, can only be walked at low tide. To the south it is usually possible to walk at all phases of the tidal cycle.

Campground Beach, Eastham

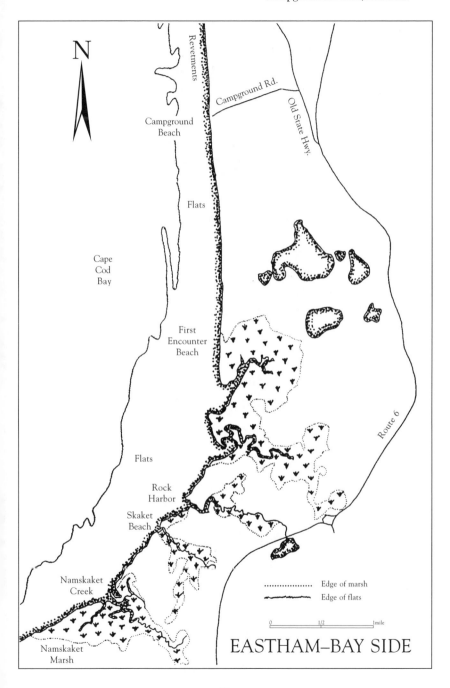

purpose? In Eastham they have probably stopped about 30 feet of sand of varying height from moving. Maybe by so doing, they do starve beaches elsewhere, although in Eastham sand has been building up at both ends of town anyway.

Just as the Incas fed maidens to their gods, some towns resolve this dilemma by requiring that revetment owners feed sand to the sea monster. In Eastham the revetment owners have been required to replenish the beach each year with an amount of sand which would add one foot to the entire area of bluff behind. I am not sure that towns impose the same requirement on beach grass planters or snow fence builders, even though if they work by holding the sand, such techniques are also "starving" adjacent properties.

In Eastham and other Cape Cod towns, no one seems to be winning. Erosion goes on, houses continue to fall into the sea, and beaches continue to disappear.

Namskaket Marsh, Brewster
Salt Marshes & "Wadlopen"

The Dutch have a sport called "wadlopen" (mud walking) in which they climb over the dike at low tide to the floor of the North Sea beyond, then attempt to walk through the mud to distant islands. The record distance is 12 miles, achieved by three tall men who arrived neck deep in water following a record low tide.

Cape Cod would be a good place to train for such competition. One cannot walk far on our shores without coming to mud in the many salt marshes created when sandy barrier beaches plug up the flow of detritus created by rivers and streams. Such marshes are considered the richest ecosystems on earth, nature's own self-perpetuating compost basins. The Cape's early settlers grazed their animals on the salt marshes, which were highly coveted farm land. Some were even known to have put snowshoe-like devices on one front leg and the opposite rear leg to keep the animals from getting stuck in the mud.

Actually, muck is a more descriptive word than mud. Both literally and figuratively it defines something rotten.

I had a healthy respect for muck before my walk around Cape Cod began. One year as Sue and I paddled furiously during low tide in the annual canoe race at Harwich's Herring River, we cut a corner too closely and ran aground. When I threw one leg over in the standard maneuver to push off, it sank up to the thigh into muck and stuck. There I was, immobile, one leg in the canoe, one stuck leg stuck in the muck. Fortunately, Sue's poling and some form of instinctive wiggling on my part eventually dislodged both my leg and the canoe.

Friends tell of having left shoes, "waders" and various other articles behind as a small price for getting out of the muck. One friend even required the rescue squad to extricate him from Maraspan Creek in Barnstable Harbor.

I first came face to face with the muck while walking with my friend, Bob Wilkinson, at Chipman's Cove, just south of Wellfleet Harbor. On the Geological Survey map the Cove looks like water, but Bob and I arrived at low tide hoping to ford it, thereby shortening the trip towards Eastham.

We timed it right. There was no water at all, just an entire cove of muck, glorious chocolate pudding-like muck. Cautioned by "the school of hard knocks," or should I say "squishy steps," we started around the Cove, constantly and cautiously trying to venture further into the cove in the hope that it would be passable. But the further we edged in, the deeper we sank. Hopes for a ten minute crossing soon became a much more time consuming walk around.

The next day I asked my friend Win Downs, who has a cottage on Chipman's Cove, "How far can you go in to that muck, Win?"

"They say all the way in some places!"

On the Cape, it is always "they" who say.

Bob was also with me when we came to Namskaket Marsh on the Orleans-Brewster border. We arrived at this marsh not at either end, but in the middle after a trip down the creek which creates it. Bob describes this encounter best in his own words:

> We could see the beach way out across the marsh so we started out with great enthusiasm—right across the marsh. Big mistake! Soon we came to a small creek which we studied for a bit before jumping over it with very little effort. After we walked a bit more, we came to anoth-

er. No problem, over we went. Soon thereafter we came to another and then another. Each one would take a little more study. We would walk in one direction and then the other looking for the best footing, always with the hope of staying dry and out of the thick mud which made up the banks of these creeks. Before you knew it, the creeks were wide enough to require a good bit of study for footing and creek bottom because we were now at a point where we had to run like crazy to get up a full head of steam to reach the elevation and velocity necessary to carry ourselves to the other side. Bear in mind that the wind was brisk enough to get your attention, even with dry clothes. One could imagine what it would be like with wet and muddy clothing. Also bear in mind that Elliott and I have approximately 105 years between us. I imagine that we were quite an amusement to anyone on the shore who might have seen us crashing along trying to get up the speed for each consecutive creek.

After what seemed like a very long time and with energy ebbing fast, we came to yet another creek. This creek—river is more like it—had to be the granddaddy of all the others and it was obvious that Tarzan wasn't going to jump it. There was nothing to do but turn back. You guessed it! Turning back meant that we had to cross all of the creeks which we had already jumped so that we could get back to the marsh's edge. From there we would have to follow the edge of the marsh, a much longer route than originally anticipated, until we could get to the end of it and onto the beach.

As we trudged back, we slowly separated, looking for the shortest route back to the edge, hopefully the one with the least amount of creek jumping. By this time we had misjudged the footing on a few jumps (there were a lot of them) enough to have gotten our feet a little wet and muddy, but all things considered, we were reasonably dry.

Before I knew it I was walking along alone with just my thoughts, no doubt thinking how nice it would be to have a hot shower and a long nap. Or, perhaps I was thinking about how nice it would be to push my walking companion into a nice deep creek to discourage him from thinking about another "weekend walk." In any event, I was out there in this god-forsaken marsh with just my thoughts when all of a sudden I found myself hip deep in mud. That is to say, one leg was hip deep in the mud while the other was lying straight out as if I had just done a split. After the pain subsided, I set myself about the task of getting myself out of a classic sinkhole. Everything had come full circle. A salt marsh is made up of meandering creeks that get wider toward the mouth, with strategically placed sinkholes, a normal enough looking surface covering up pure deep mud. The lessons of my Wellfleet youth had been revisited.

So, there I was wandering around Namskaket Marsh in a Northwest wind, now covered with mud including a fair bit inside my boot, with my left side soaking wet.

Finally, we made it back to the marsh, right back to where we started, still a long way from our final destination in West Brewster. As we had been dropped off by Elliott's wife we had no choice but to carry on. We did and we made it. Since then I've made a few more crazy treks with Elliott. Each time somewhere along the way I say to myself that enough is enough and I promise my failing legs that this is the last time, really believing it; but some time goes by and I get another call from my walking buddy saying "I'm planning this walk, do you want to join me?" What do you think I say?

The Sand Flats of Brewster
My "Home" Beach

The sand flats of Cape Cod Bay, which extend at low tide from South Wellfleet to Sandy Neck in Barnstable, are a constantly changing maze of sandbars, shallow pools, and irregularly interspersed channels. They provide an easy way for a walker to avoid the marshes, muck, and private property on the shore.

On a sunny summer day, the flats are a beehive of activity, covered with blanket-based sun worshippers, sandbar hikers, swimmers, shell-fishers and shell collectors, and in places four-wheel-drive vehicles coming onto the flats from town landings in the adjacent towns. Although people are everywhere, as one gets further out on the flats, there is a special sense of space and solitude. Eventually, most of the figures become distant and appear small enough to be reminiscent of the toy figures in a child's sandbox.

I get a profound sense of release on the flats, as if I am playing hooky; cheating the system. This sense is partly imaginary, but it is also partly real: the further out one goes onto the flats, the further the indelible marks of man fade into the distance. Although I understand the need for people to make changes in the landscape to feed, house, and employ themselves—and to consume resources, including land in the process—I do not recall many places where this has improved the view. Certainly I have never seen a utility pole as lovely as a tree, or a parking lot as beautiful as a field. On the flats there are no trees or utility poles, no fields or parking lots, but there is a natural eraser—the tides—which sweeps over them twice a day and returns them to their pristine state.

I am not a shell-fisher, a sunbather, or even a swimmer—except when swimming is the only method of getting somewhere I want to be. For me the joy of the flats is in being there and walking. And the sport of walking is in the seeing, the thinking, and the talking when with friends. Our society underestimates the importance of thinking and the need to find the right times and places for it, and overestimates talking. While walking with friends can be more fun, walking alone can be far more productive.

One beautiful September morning I set out at 7:05 a.m. from the town landing at Dyer-Prince Road in Eastham, intent on walking approximately eight miles to Sea Street Beach in Dennis during one low tide, traveling much of the way along the outer bar.

At least one swim and considerable wading would slow me down along the way. With low tide due at 8:50 a.m., I expected to have slightly over three hours to make it. I carried only a zip-lock bag of raisins and two water bottles of Gatorade.

The outer bar is a continuous sandbar which defines much of the seaward edge of the sand flats of Cape Cod Bay: the most distinctive and highest of the many bars throughout the flats. In contrast to the miles of shallows on its inside, the water level drops off sharply on the bar's outer side.

It is not clear why the outer bar is there. It may be a portion of a sunken shoreline curving around from Dennis to the Billingsgate Shoals in Wellfleet, dating back to the glacial age which formed Cape Cod. Or it may be a product of the continuing erosion process, lighter residue carried further after the heavier sediment dropped out at Sandy Neck in Barnstable and Provincetown.

Most people understand that some high tides are higher than others. Less thought is given to the fact that some low tides are

The Sand Flats of Brewster

much lower than others: indeed three and a half feet lower in Cape Cod Bay. There is a considerable difference between walking on sandbars and wading, and even more difference between wading in two inches of water and wading in three feet of water, so I waited about a month for a convenient low, low tide. That morning's low, although six inches above that of two days before, was still very near the low-end of the regular bimonthly cycle. Just two days later, low tide would be a full foot higher.

The water clears out of the inside elbow of Cape Cod, where I started that day, well before low tide, so at the beginning there was little difficulty.

The flats were alive as they can only be in the early morning. Leaving shore I walked through a fountain of geysers from spurting clams, some reaching a height of four feet or more. I have never seen more gulls and terns. What a racket they made! I knew from living on the herring run in Brewster, that gulls are loudest in the morning. I wondered what they were saying: "I'm hungry," "beautiful day," or perhaps "I've got that one." Perhaps, like humans, it is a full stomach that quiets them down. In any event, the wild creatures of the early morning flats seem to regard it as their time. The first human I saw was a solitary jogger at Skaket Beach in Orleans, a quarter of a mile or so inland. There were no human footprints disrupting the pristine beauty of the ripples in the sand.

I passed the dead tree branches inserted in the mud to denote Rock Harbor channel in water only halfway up to my knees, and headed directly for Ocean Edge in Brewster, where scouting missions had indicated I would have to come closest to shore. Even well out on the flats there is a distinctive channel and flow for each creek—Rock Creek, Little Namskaket Creek, and Namskaket Creek. Fortunately, none represents a serious wade at low tide.

The nature of the flats changes in Brewster. There are more patches of seaweed and more small rocks. Even the sand is different, darker and intermixed with more natural rotting vegetation.

On this day the wildlife and birds changed, too, as I approached the shore at Ocean Edge. A razor clam slithered along a shallow, squirting water like a clam, in this case perpendicular to the sand, seemingly to propel its movement.

I walked as close as I could to a great blue heron. It stood first like a lone sentinel, then walked through the shallows like some stiff-legged clown on stilts. As it soared and landed, having taken flight to put another hundred feet or so between us, it looked with wings spread like a stealth bomber or SST coming in for a landing.

Flocks of sanderlings, plover look-alikes, had for the most part replaced the gulls and terns. They were delightful companions as I walked the Cape's shores, usually scurrying two or three times to keep their distance ahead, then taking flight to circle around behind to their original spot. Sue or I would often circle around them to sit silently in ambush on the beach ahead, hoping to photograph them as the other herded them into the trap. They never fell for it.

Off Ellis Landing I headed back out towards the outer bar, visible a mile or so away. A little more than an hour into the walk, I engaged in my first human interaction with several members of the Ellis family out tending the clams in their aquaculture pens. They rewarded with me with the report of a newly discovered hand hewn wooden pipe which had recently been exposed on the beach. The pipe was undoubtedly once used to carry water to some enterprise on the shore. Something to explore on another day.

Erosion comes and erosion goes; erosion giveth and erosion taketh away. A revetment was built last year beneath the cottages at Ellis Landing after a winter storm washed away a few feet of the bank. Perhaps the same storm had helped expose the pipe.

I was a little apprehensive about the walk back to the outer bar. Scouting missions, performed during higher low tides, had necessitated considerable waist deep wading without even crossing what appeared to be the deepest water. Soon, a new difficulty arose, the leading clouds of a cold front began to obscure the sun. When the sun was out, its reflection on the light colored sand made the entire bay look welcoming and empty—easy walking. But whenever the sun went behind a cloud, as little as an inch of water looked mysteriously dark and uninviting. Fortunately, I could always see the whites of the birds and waves breaking on my target—the outer bar—so the changing appearance of the area in between never altered the goal, even though the coming and going of clouds did make it more difficult to pick the best route.

I have read that the flats in this portion of Brewster are the widest in North America. I am not sure that I believe it. Brewster has less shoreline than any other Cape Cod town—just 5.7 miles—but much of it is publicly owned and the private owners and walkers seem respectful of each other throughout. At low tide the outer bar encloses Brewster's entire shore front in a community sandbox up to a mile and a half wide.

Caution must be exercised while walking the flats. There are places where the outer bar is separated from land by pools of varying depths. As the incoming tide returns, it is possible to find oneself separated from shore by irregular and rapidly deepening water. On a normal day, the tide rises about a foot and a half an hour.

My walk back out to the outer bar went well. After two short waist deep wades I reached it off Breakwater Beach in time to examine three large stakes, probably the remnants of an abandoned fish weir. The bottom nine feet of the remaining stakes were dark, but from a distance the tops appeared to be painted white. They were not—closer inspection indicated that the two colors portray the height of the tide, the differing colors the varying results of weathering in the water and air.

I turned along the outer bar, more than a mile out to sea, knowing that it was now a straight line walk to where it rejoined the shore at Sea Street Beach in Dennis. Walking the outer water line, I could see that each wave now advanced a little further than its predecessor. As the two hour mark of my walk approached, the tide had changed.

My legs were becoming a little rubbery and one serious obstacle remained, a channel through the outer bar off Brewster Park which was deep enough to require a swim.

For the first time I considered what could go wrong. There were not many people around, my legs were tired, and there was a very strong wind blowing offshore from the southwest. The wind had probably helped the tide empty the water from the bay, but represented a difficult head wind for a walker, and would certainly not help me get where I wanted to go if it became necessary to head for shore through water well after low tide. What would happen if, walking alone, I got a serious leg cramp?

Compared to the rest of the flats, portions of the outer bar are often a beehive of activity. On a sunny summer day boats stream along the outer edge, some beaching on the bar so that the occupants can utilize the privacy for a picnic or swim.

Old time Brewsterites tell me that during prohibition the privacy of the bar was utilized for a very different purpose—unloading the large vessels of rum-runners to transfer the cargo into alternative means of bringing it to shore. Every now and then something would go wrong, leaving free booze available on the bar for anyone prepared to make the trip out at low tide to get it.

My spirits rose as the number of humans nearby increased, indicating that I was approaching the channel. First were four men with rakes and bags, clamming on the outer bar as I neared the channel. It was the easiest clamming I have ever seen, clams literally poking through the surface as I walked along. The men did not even have to bend over or break the surface with their rakes to fill their bags.

Both sides of the channel were lined with fishermen in waders casting for bass. Several had arrived in a canoe, having paddled out through the lagoon which spreads both ways inside the channel between the outer bar and the shore.

I had planned to swim the channel, but the lack of a life preserver, the head wind, and my rubbery legs all led me to conclude that discretion was the better part of valor. A detour requiring two chest deep wades across both ends of the lagoon took me around. Physically, this was the most difficult portion of the walk. As I waded, carrying my pack overhead, the bottom was extremely erratic. I never knew whether a step would bottom out in water slightly above my waist or up to my neck. I proceeded with extreme caution. The sand on the far side of the lagoon was equally erratic—looser, wetter, and mounded a foot or more—very difficult walking for tired legs. I considered whether to head in to Paine's Creek Beach, my planned bail-out point, but I felt too close to the destination to quit.

As I waded back out across the far side of the lagoon, an attractive young woman in a black bathing suit approached along the outer bar from Dennis. Black bathing suits are a family joke in our house. Noticing once while walking in mixed company that there are a disproportionately large number of them, and knowing that black absorbs the sun and thereby warms, I innocently asked whether women wear them to warm their bodies. My companions howled with laughter. "No," I was told, "they make you look slimmer."

The woman approaching could have worn a suit of any color. She looked like a mirage on a tropical island—deserted except for the two of us. As I waded out of the water, she passed me, stopping to turn around, gaze at the beautiful stretch of sand behind her, and say, "It is gorgeous and it is all yours." She did not turn around to walk with me. I wondered if she had walked all of the way out from Dennis just to swim in the lagoon, or if she was on her way to Eastham.

The remaining stretch of the outer bar was firm, flat, several hundred yards wide, and still about two feet above the water. Even on my wobbly legs and with the wind blowing hard enough to blast dry sand into my face, it was the easiest walking since the beginning of the day's journey. The last mile or two were pleasant and uneventful, as the bar rejoined with the shore at Dennis, disrupted only by one more channel to wade, that of Quivett Creek. I reached the parking lot at Sea Street Beach at 10:15 a.m., exactly as planned, tired but satisfied.

The ocean comes closest to my home in the area where this walk ended. It is my "home" beach. On a day when I have nothing else to do, I will cajole a family member into driving me to Sea Street Beach so I can walk home via Paine's Creek Beach in Brewster, the closest landing to my home. There are a variety of choices for walking on the way: out on the outer bar or flats; along the sandy beaches of much of the shoreline; or in among the varying attractions which lie just inland. There are two conservation areas (Crow's Pasture in Dennis and Wing's Island in Brewster), two creeks (Quivett in Dennis and Stony Brook in Brewster), as well as the two salt marshes which surround the creeks. The marshes are particularly alive in the spring when the herring are running. A walk at that time of year almost deludes me into believing that I am on a bird-watching safari in Africa, but I am never actually more than a mile or two away from home.

My regular home beach walk goes east from Paine's Creek Beach down the Brewster shoreline to Breakwater Beach. I walk it several times a week in all seasons, at all times of day and night, and in all forms of weather. The other regular walkers are a realtor, a school principal, a landscaper, a lawyer, and another banker. Occasionally we are accompanied by a rookie or two, various family members, or, in the off-season, up to five dogs. All but the dogs find the conversation too intense.

We talk about everything on our walks, but about politics and schools the most. We never agree on anything but the facts. Nothing anyone has said has ever changed anyone's mind.

The Sand Flats of Brewster

The beach is always the same, which is to say it is always different.

In the summer my home beach is typical of the Cape Cod shoreline, a series of town beaches crowded with happy vacationers mixed in-between the private beaches of several condominium projects and individual homeowners. There are few "no trespassing" signs and most private owners and beach walkers are respectful and tolerant of the other.

By late fall, the people are entirely gone, as is the color provided by the beach grass.

I particularly remember the walk one cool November evening after the beginning of the winter's abnormally high tides. The moon was full, with an enormous ring around it. As jets made their way overhead to Europe, the smoke trail of one of them turned the ring into the international version of a stop sign, a circle with a line through it. It was yet another reminder of the intrusion of technology on nature.

That night the full moon made it easy to see objects on the beach, a stark contrast to the recent series of dark, foggy nights which had constantly reminded me, as I stumbled over the rock groins, that my night vision does not improve with age. My companions have since given me a hat with a light attached to the front which I can turn on as I maneuver over the rocks. I wonder if they are trying to tell me something.

It had been a little more than a week since we had last walked the beach. The north winds had produced much change since our last viewing. The beach had quickly put on its winter coat. The flat at the top was gone, it being all one slope from bluff to waterline, and the terrain had been scoured (or pounded) into a hard surface. There was more sand on the west side of the groins and less on the east. Friends with a home two miles away, where there are no groins, said that they had lost eight feet of beach.

The winter of 1995-96 was not an easy one for the Cape Cod Bay shoreline. By the time we walked it in December following the highest of the unusual "spring" tides, the erosion was the worst on the bay in years. Cottages perched close to the top of the bluff, from which large chunks of once solid land had collapsed.

Since it washed in a few years ago, one of our measures of the changing beach has been a large wooden "beam" comprised of several boards bolted together. All told, the beam, which rests on the high tide line, is approximately ten feet in circumference and thirty feet long. It moves with the high tides of the winter, but only in winter. This year it had been lifted by the sea from the leeward side of the groin and rested in a considerably higher spot on the other side, having moved upward and into the commonly prevailing current along this shore. We marveled at how the sea could have muscled and maneuvered this particular move.

By spring the scene was returning to normal—if there is any such thing on the shore. Loose sand was again building a plateau just below the bluff, and the sand was softer throughout the entire slope of the beach.

This is the beach where I view horseshoe crabs mating, birds trapped in fishing nets, living and dead birds, fish and animals of many varieties, as well as a myriad of other sights and scenes—common and uncommon. This is what I believe having a home beach is all about.

WALKING IN WEST BREWSTER AND EAST DENNIS

The area between Cold Storage Beach in Dennis and Paine's Creek Beach in Brewster offers many attractions: the outer bar, the sand flats, a sandy beach, and the salt marshes and streams behind the beach.

One end of the outer bar leaves the shore just east of Sea Street Beach in Dennis. This bar is over a mile out by the time it passes Paine's Creek Beach in Brewster. There are pools of varying depths inside of this bar, so anyone attempting to visit it should be well informed of the timing of the tides and have a planned route back in. At low tide there are several relatively dry routes out, including one at Paine's Creek Beach.

The vast expanse of the sand flats inside of the bar also provides

The Sand Flats of Brewster

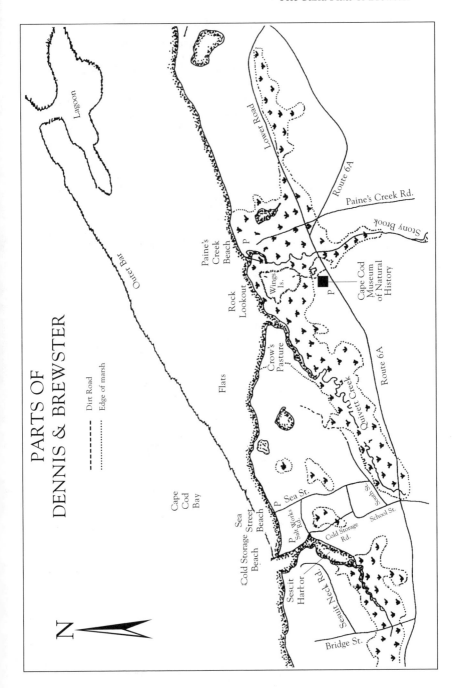

wonderful walking, wading, and exploring at low tide.

There are also several routes along the shore and slightly inland.

The sandy beach is broken only by the mouths of Quivett Creek and Paine's Creek. Portions of this beach are privately owned, but signs permit passage for through walkers.

Inland of this sandy beach are the salt marshes surrounding the two creeks, providing yet another "shoreline," some of which also crosses private property. Crow's Pasture conservation area in Dennis and the Wing's Island area, maintained by the Cape Cod Museum of Natural History for the Town of Brewster, provide access to explore this area.

This entire area played a prominent role in the history of Cape Cod. Plaques commemorating the first saltworks on Salt Works Road on Quivett Neck and the Shiverick Shipworks on Sesuit Neck Road are not far apart in East Dennis, although on opposite sides of Sesuit Creek. East Dennis village remains unspoiled and is a very picturesque walk. Wing's Island in Brewster was populated by Native Americans and was possibly the home of Brewster's first settler of European stock.

Barnstable Harbor
Mixed Adventure

What natural daylight there was on this damp, October Saturday had already begun to fade as my companion, Peter Norton, and I left the trail on the south side of Sandy Neck and began to head across the salt marsh toward the waters of Barnstable Harbor.

We had spent the afternoon exploring Sandy Neck, which offers a choice of walking alternatives—an open sandy beach on the outer shore or a path along the salt marsh on the inner side. In between, the land mass is large enough to be home to a number of scattered hunting camps, a colony of beach cottages, and various other vestiges of human activity. It is only accessible by foot or four-wheel-drive vehicle and, in places, only at low tide.

Peter and I faced a variety of obstacles before the day was complete. First, the marsh had to be crossed—a part of the Great Barnstable Marsh, the largest on Cape Cod, divided by numerous mucky creeks which would have to be jumped or circumvented. Next we would have to cross the harbor itself, a maze of hummocks, channels, and sand flats, some covered by varying and unknown amounts of water. Before done this day's leg would encompass all of the adventure forms characteristic of my walk around Cape Cod.

The harbor crossing had been planned for just west of the boat channel into Barnstable Village. Maps and conversations with friends had indicated that this route would minimize the need to swim, perhaps eliminating it entirely. Low tide was scheduled for 5:30 pm, providing less water to cope with and, soon thereafter, a tidal current flowing into the harbor, not out.

Progress was slower than anticipated.

Since help was a long way off if one of us got stuck in the marsh, we proceeded cautiously, often moving sideways as much as forward to avoid creeks and the soft ooze that continually blocked the path.

After finally reaching the edge of the marsh, we faced the next obstacle, a drop of about six feet to the floor of the harbor. It would have been a fun slide, but again we might have encountered muck of unknown depth at the bottom if we slid straight down. We walked along the edge for a short distance before discovering a more gradual route to the bottom.

Peter is a large and amiable man. His presence on this leg illustrated his adventuresome nature. I had told him that "there should be little or no swimming" but after a short walk on the flats at the bottom, we came to our next surprise, a channel which was clearly over our heads. Peter could not resist the temptation to point out the inaccuracy of my forecast, "We have hardly started, but already I can see water everywhere."

We were dressed in life jackets and bathing suits, as I had been for most of my journey because I never was sure when I would have to swim. But, hopeful of a drier route, once again we edged sideways. It must have been stubbornness; we were already wet from an afternoon of drizzle and exercise.

Soon we approached two shellfishermen, working beside their small boat. One, who had dug himself in waist deep, appeared from a distance to be a dwarf. Shell-fishers are like security blankets on the marsh. They always made me think, perhaps foolishly, that "this cannot be too crazy a place to be."

"What is the driest route to Barnstable Harbor?" I asked. We were still the better part of a mile away.

"Sure enough," one answered as he gestured toward a

number of points of varying size and distance away, "if you go over there to the end of that point you won't have to swim."

We scrambled back up on a hummock, walked to the end ("the point?"), scrambled down, and "sure enough" waded across the channel.

By now it was past low tide and dark enough to make it difficult to visualize how deep the water was, even at one's feet. We started to wade, but soon again came to water over our heads. Well behind schedule, we swam without hesitation, constantly attempting to touch bottom. Reaching shallower water, we waded a considerable distance in water ranging from our knees to our necks. Finally, we realized that we were only several hundred yards from shore beyond the mooring basin for the Barnstable Yacht Club. Looking at the size of the boats, we realized that the water would be deep enough to require a swim. Shortly after we started, an unseen tidal current began moving us sideways faster than we were capable of swimming forward, making it likely that the duration of this swim would be longer than either of us had anticipated or hoped.

I was never frightened while swimming, but was always anxious to be done. There was too much that could go wrong: power boats, swift currents, and, as summer gave way to fall, the possibility of colder water. Hence, I always tried to maintain a margin of error by resting whenever possible. I attempted to grab onto one of the moored boats. In the current it took four tries before I correctly gauged the ratio of sideways movement to forward progress, grabbed a rudder, and paused to reflect.

About halfway to shore, the current stopped as suddenly and unseen as it had started. We finished the swim in total darkness, there being no moon or stars that night, and arrived at the yacht club just after the lights were turned off for the night.

With dismay we realized that we had not checked our bearings carefully enough when we parked our car prior to the day's excursion at the town landing at the end of appropriately named Rendezvous Lane. In the dark we did not know which way to go along the shore to retrieve the car. We started toward Route 6A. The occupants of the first car that passed us were visibly startled to encounter two men walking in the drizzle after dark, clad in bathing suits and life preservers and carrying a water-tight yellow bag. Either from confusion or a desire to get us out of the neighborhood, they told us the wrong way to go on Route 6A. After dutifully following their instructions for a half mile or so, we realized our mistake and backtracked to Rendezvous Lane and the car. I quickly called home on the car phone, fearful that by now Sue would had called the Coast Guard.

If I had it to do over again, I would have planned this segment of the trip a little differently.

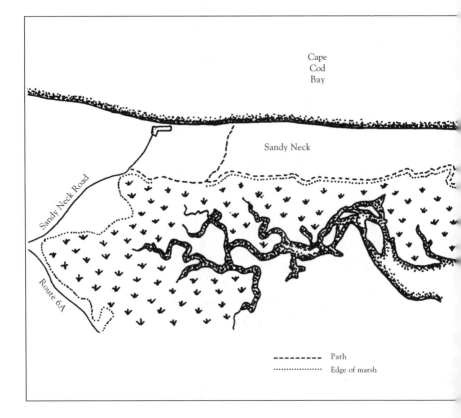

WALKING SANDY NECK IN BARNSTABLE

Sandy Neck is a barrier beach peninsula over six miles long, built up by eroding sand traveling along the coast.

A marsh trail along the south side provides excellent views of the 8,000-acre Great Barnstable Marsh and Barnstable Harbor. The trail passes several privately owned hunting and fishing camps. Portions of this trail are under water at high tide, but usually there is a way around.

A beach trail along the outer side provides typical beach walking. There are a scenic lighthouse and village at the end of the peninsula; a reward for the hiker capable of the 12-mile round trip. The cottages are on private property. Privacy rights should be respected.

Four short trails cross the peninsula from marsh to beach at strategic points.

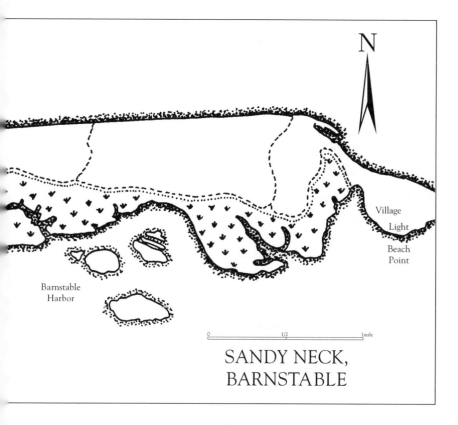

SANDY NECK, BARNSTABLE

The Cape Cod Canal
Tides

The Cape Cod Canal, completed in 1914, is unquestionably one of the most substantial changes that man has made to Cape Cod. Enough material was excavated to build the dike to Mashnee and Hog "islands." How the world would squawk if the Army Corps of Engineers proposed to build it now.

The Canal is a busy place. The fishing is good, the walking and biking are good, and it is the only way to go between Buzzards and Cape Cod Bays in a boat. But, the current in the canal is one of the strongest in the region, seemingly always in a hurry to get from one end to the other.

Two different tidal flows meet at the canal. The rise of the water from low to high tide (tidal range) is three feet in Buzzards Bay. However, at Cape Cod Bay on the other end, it is nine feet. It doesn't take a scientific genius to realize that a considerable amount of water must flow back and forth through the canal in order to compensate for that difference.

The water in the Canal is particularly interesting to view when the tide is flowing in one direction—literally pouring the water from one end to the other—and the wind is blowing in the opposite direction. While the wind attempts to blow the tops of the waves in one direction, the current moves the bottoms in the other. As a result the waves struggle to stand still. When a boat's wake comes along to help them out of their confusion, the wake seems to head for the side with more than the normal enthusiasm.

"Why is the tidal range three feet in Buzzard's Bay and nine feet in Cape Cod Bay only several miles away?" I wondered as I passed by.

It was not an easy quest to answer this question.

We learn in schools at an early age that tides are caused by the gravitational pull of the moon. Tides are high when the moon is overhead and underfoot (on the other side of the earth), and low when it is at right angles. There is a tide in every body of water, even the smallest pond, but some are just too small to notice.

Sounds easy, but why does high tide in Buzzards Bay occur several hours before it does in Cape Cod Bay? Seemingly the moon passes over and under both at approximately the same time? Now I had two questions to answer.

I looked in every book I could find about Cape Cod and then at several books solely about tides. Unfortunately, most of the latter were written by physicists and frequently refer to determinants such as "hydraulics" and "oscillation" among other causes of tidal flow. They even have sections written in algebraic Greek which give you the formulas to measure the volume of each type of flow. Ultimately, I learned the most from the World Book and the special edition of Rachel Carson's The Sea Around Us written for young readers—Carson's actually contains more information and better pictures. Maybe I am not as old as I thought. Or, perhaps as we grow older, we should start reading books for children again.

In order to spare you many pages of complex elucidation, here is my best shot at a brief, simple explanation.

Bodies of water vary in shape and depth, among other things. Water takes time to move. Friction with the sides and the bottom slows down and reflects (reverses) the basic tidal flow, which in the open ocean can move as rapidly as a thousand miles-an-hour, the speed at which the moon passes over the earth's surface. As a result, for example, high tide in the innermost part of Pleasant Bay is about three hours later than at the mouth, while the inner range is three feet compared to seven at the mouth. With all those nooks and crannies bouncing the basic tidal flow around, it takes the peak tidal flow that much time to traverse the short distance of Pleasant Bay, and not all of the water ever gets in or out. The process gets even more complicated in a place such as Nantucket Sound, which the tidal flow enters through three inlets on three different schedules. At one end of Nantucket Sound in Chatham the tidal range is about seven feet and high tide is over an hour later than at the other end, Nobska Light in Woods Hole, where the range is only about one and a half feet.

Probably of most interest to Cape Codders, however, are the unusually high and low tides inappropriately known as "spring tides" even though they occur twice a month year-round. These highs are responsible for considerable erosion damage, hence they are the tides most feared by owners worried about losing their property to the sea.

Why do spring tides occur, and why are they most extreme in winter?

Simply put, the sun also has a gravitational pull. "Spring" tides occur not when the moon is full, as many believe, but when the moon and the sun are aligned so that their gravitational pulls work together. "Neap" tides, with the smallest range between high and low, occur when the sun and the moon are in perpendicular positions in relationship to each other.

One final comment—not all "spring" tides are the same either. For one thing, tides are not really highest when the moon is directly overhead. Gravitational pull is stronger sideways than directly upwards. Water, like wagons, is easier to pull sideways than lift up. For this reason the more extreme "spring" tides tend to be in the winter months. That is when the sun is furthest to the south, maximizing the sideways pulling effect in the northern hemisphere. Furthermore, tides are highest when the moon, which rotates around the planet on an elliptical course, is closest to the earth (at the perigee).

Samuel Eliot Morison wrote in Spring Tides that extreme tides, "for some scientific reason unknown to me, always come within an hour of noon or midnight." The reason would seem to be that these are the only times when the sun is directly overhead or underfoot, with its pull maximized, and hence the only time when the sun and the moon can be aligned with a maximum tidal pull.

The timing of the most severe spring tides is thus easy to predict. They always occur in the winter months when the moon is overhead at noon or midnight. Over the years, this information has been of use in a variety of ways. Early ship builders, including those on Sesuit Creek in Dennis, floated their boats out on each year's highest spring tides. Today, the "need to know" is often for less industrious reasons. For instance, it is the time to go to Chatham if you want to see a house fall into the sea. Isn't it convenient to be able to do it at noon? I once canoed around Pochet Island in Orleans, which is only truly an

---

WALKING THE CAPE COD CANAL

   Army Corps of Engineers maps are available at several visiting areas for walkers at the Cape Cod Canal.
   Paved service roads (closed to vehicles) along both sides of the canal, are utilized by many walkers, bikers, and skateboarders. The entire distance provides an excellent view of the unique water flows in the canal. The "people viewing" is also superb.
   The Bournedale Interpretive Trail on the north side between the Herring Run Recreation Area and the Bourne Scenic Park offers an interesting alternative for those seeking relief from the activity along the service roads. Although the trail goes through a narrow band of vegetation between the service road and the highway which is not large enough to eliminate traffic noise, the foliage is thick enough to provide the sensation of walking through the woods.
   A mile east of the Canal in Sandwich Harbor there is a very scenic boardwalk rebuilt through public subscription after destruction in a storm. After crossing the boardwalk, it is possible to view the abandoned channels through the marsh and the site which once served the Sandwich Glass Works

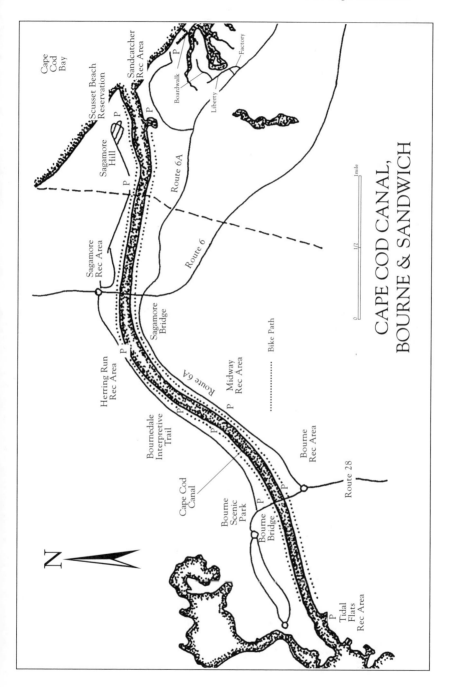

island during "spring" highs. After canoeing over a salt marsh, we paddled down what is normally a dirt road. It can be somewhat disorienting to look down from a canoe and observe car ruts.

But, go to the Canal if you want to observe tidal flows in action, and you cannot make it to the Bay of Fundy in Nova Scotia where the tidal range is 50 feet, and the tidal "bore"—a literal wall of water leading the incoming tide—is six feet high. This makes me wonder why there is a tidal bore in the Bay of Fundy, but none in Pleasant Bay. We have already established that in Pleasant Bay not all of the water has time to make it in or out. Would not a bore help? The answer has something to do with differing shapes of bodies of water and differing distances from the Continental Shelf.

There appears to be bore on Cape Cod Bay. After low tide on the flattest part of Provincetown Harbor, the visible returning wall of water is at least one half of an inch high

Buzzards Bay
Family Walking

At the beginning of my walk around the Cape, I concentrated first on the areas closest to my home in Brewster, gradually moving further and further out.

As I reached Buzzards Bay, Sue began to join me more regularly. We selected a walk of five to eight miles, then parked one car at each end.

Our walking styles are very different. Sue's comfortable pace is slower than mine; she likes to smell the roses. As one friend puts it, "there are bird watchers and there are hikers." Initially, I walked ahead of her, going slower than I otherwise would have, while she trailed, going faster than she would have. If I slowed down, she slowed down.

From Sue's perspective, our walking styles are even more diverse than I have indicated. Her description of the situation follows:

> As would be expected, there were many situations or problems on our walks which had to be coped with or ignored in order to keep the walk going—some petty, others more complicated. For instance there was the constant build up of sand in our sneakers, the discomfort of long stretches of wet, muddy feet, the aching frame, or the occasional crisis like the dense cloud of 'no-see-ems'

after dark in Chatham. These were all relatively easy to put up with. Elliott's and my different walking styles were not so easy to endure...

Because Elliott had an overall plan in his head with a carefully arranged time schedule that fit with the tide, the beach, and the time of day or night, his modus operandi was "full steam ahead." I, on the other hand, was along for the ride. Each beach was new territory to be explored. I preferred to watch what was going on and to take pictures—to make sure that I didn't miss anything. His goal was to cover territory, compare each retaining wall with the last one, and file it away in his mind. We drove each other crazy.

Once in Provincetown we were planning to spend all day walking, so I packed a lunch. Come lunch time, his desire was to eat while walking, mine was to sit down for a picnic. The compromise was that I sat until I had finished my sandwich while he stood eating and looking longingly in the direction which we were headed.

At Scraggy Neck on Buzzards Bay we were walking at dusk. All the conditions were present for a spectacular sunset. The sun made a golden path on the water which, combined with the waves and the rocks, was a real show-stopper. To add to this, there were two boats on the horizon headed in opposite directions, one a large steamer and the other a three-masted sailboat under full sail. The boats appeared as if their silhouettes might even cross in front of the sinking sun. I stood there with my tongue hanging out. It was a while before I realized that Elliott was a good distance down the shore, striding earnestly on.

We learned to compromise. I stopped taking pictures of every cormorant we saw, and he tried to moderate his pace. Each style has its purpose and each has its drawbacks. Fortunately, we were able to cope with these, sometimes more gracefully than others.

I must admit that, unlike my wife, I got into walking as a form of cardiovascular exercise. After a somewhat sedentary phase of my life, I became interested in regular exercise. Part of my indoctrination was that the heart requires a steady workout. If you stop to rest—as in the game of Monopoly—you have "to go back to go," in other words start again, to derive any benefit.

As further justification for maintaining a very steady pace, I rationalize my preference by reminding myself that all parts of the view are equally deserving of time.

Our differences in style were undoubtedly exacerbated by the complexity of the Buzzard's Bay shoreline which made it far more difficult to predict the difficulty and duration of a walk. This shoreline is dramatically different from the rest of Cape Cod. It is rocky and irregular, having not yet been eroded almost entirely into straight sandy beaches like most of the other "sides" of the Cape.

On Buzzards Bay there are big rocks, small rocks, and in-between rocks. There are also relatively flat rocky "beaches" and steep rocky banks. Of more importance to walkers, depending on where and when one is walking, there are dry rocks and wet rocks, the latter of which are usually synonymous with slippery rocks.

Rocks do not take the fun out of walking. But, they do vary the footing and require more concentration. By doing so, they significantly slow down progress. Walking on rocks is more work and less play.

Furthermore, although Buzzards Bay—like the rest of the Cape—has its beaches, its marshes, and its streams, it is almost never straight, almost always going out to a point or into an inlet. This shoreline is constantly changing. Elsewhere it was easy to estimate mileage with a map and a ruler. On Buzzards Bay, such a technique almost invariably underestimated the distance around coves and points—sometimes dramatically. Not every little bump and wiggle is reflected on even the best maps of Buzzards Bay.

Cumulatively, the slower pace of traversing rocks and the tendency to underestimate distance made it more difficult to decide where to park the car at the end of each day's walk and made me feel less in control once each walk began. We were almost always behind schedule, and I am allergic to being late.

Partly for this reason, even in our first few walks on Buzzards Bay, Sue and I would walk ourselves into sync, not out of it. The closer we got to the end of each walk, the less I worried about the uncertainty ahead. But, after a few such legs we reached an accommodation which greatly increased our mutual enjoyment. We would park one car at the middle of the intended walk, not the end, then walk the first half in Sue's style. While I cranked up the accelerator for the second half, she would go off and buy a picnic lunch. Usually she would be happily sun bathing when I arrived at the final destination.

I also bought a camera to see if that would slow me down a little. But, we found out that we take pictures differently, too. To me, pictures are like written notes; you try to get as much as you can into each one—for example, two or three objects with contrasting styles. To Sue, photographs are works of art. She looks at every object from every angle, trying to record it in the most beautiful possible way.

Now that we have learned to accommodate each other's walking styles, it appears we may have to walk around the Cape yet again, in order to get our photographic acts together.

Patuisset Point, Bourne
Neighborhoods & Houses

For a walker from the Lower Cape, it is for some reason a surprise to discover that much of Bourne is just like the rest of the Cape. It was, indeed, the Cape's first fashionable summer resort—the location of Grover Cleveland's summer White House in the 1890's. There are many miles of attractive shoreline in the town, and more boats than in any other community on Cape Cod.

But what really distinguishes the shoreline in Bourne is that much of it was developed for summer use prior to the invention of the automobile, television and other technological advances which now dominate our lives. As a result, both the neighborhoods and the houses are different from those typically built today. The houses are older, often larger, and frequently closer together. Without any easy way of traveling far from home, most of the houses look like they were made for living in, as well as for looking out of. The porches are bigger and located in the front— as if the residents wanted to enjoy their neighbors, not to avoid them. Indeed, back then they built neighborhoods, not just houses.

There are seawalls everywhere in Bourne, an integral part of many shore-front neighborhoods. Many add character to the neighborhood, giving the appearance that they have successfully fulfilled their function for many years. In some places there are triple concrete seawalls, thus equating the bay with a World War II armored division when it comes to potential threats which must be defended against.

Although I would not be so presumptuous as to rate neighborhoods after having walked by them once, part of the joy of walking is in imagining what it would be like to live in the neighborhood which you are currently passing through. If there is any one place which I remember as symbolizing the Buzzards Bay shoreline it is Patuisset Point on Hen Cove in the Pocasset section of Bourne. The houses are old, comfortably sized—not too big and not too small, and close together—but not too close together. I suspect that when they were built early this century, the owners viewed this as the best place in the world. Perhaps they still do.

The homes all face Circuit Road which circles the Point between the houses and the water. There are no houses on the water side of the road, just a small seawall. As a result, the neighbors "share" both the road and the waterfront. It was a nice fall afternoon as Sue and I strolled through. Many residents were out enjoying the weather, each other, and the neighborhood. Kids skateboarded and engaged in other play, while their elders walked and talked.

The view is attractive in every direction. To the left is Hen Cove, a part of Red Brook Harbor, with Scraggy Neck beyond. To the right is Pocasset Harbor, with Wings Neck behind. Boats fill the harbors on both sides. Ahead is Bassett's Island, over which the sun sets, with the main body of Buzzards Bay visible beyond.

The Cape would have done well to have zoned some areas exclusively for seasonal operation long ago while it still could have. The seasonal cottage colonies which once characterized the shore were wonderful neighborhoods. The very lack of amenities made the cottages memorable places to be. I will never forget such boyhood experiences as lying in a bed on a stormy night in a cottage which my grandparents once owned on the other side of Buzzards Bay, listening to the rain pounding on the bare boards above and beside me. The cellar-less and "washaway" nature of the construction made the stakes in the battle against erosion far lower. The coming and going of everyone each summer made every year into a reunion of special friends. It is so easy to see only the good side of everyone when you are only with them for a summer of vacation play. And, the neighborhoods themselves placed far less burden on the resources of both the environment and the communities, than do the year around houses which have replaced many of them.

Our spirits were so high after our visit that Sue and I paused to ponder whether or not to swim across Hen Cove, even though it was closer to high tide than low and further across than we normally attempted to swim. It looked to be about a mile around by land or a quarter of that by sea. The channel markers, if accurate, indicated that the wades on each side were likely to be as long as the actual swim. It was one of those beautiful fall afternoons which make you feel liberated from all the cares of the world. Sue, as usual, voted that we swim.

A few residents out walking gathered as spectators. They seemed astonished that anyone would swim the cove at all, let alone in mid fall. No one ever seems to regard their own cove as something to swim across. Maybe it is because, unlike us, they have no reason to get to the other side. Having no habit of swimming around in the middle of the cove, they were little help when we asked about the accuracy of the channel markers. The swim was longer than anticipated before we touched bottom on the far side, but a perfect departure from a very pleasant place.

Patuisset Point was once an island. To the northeast it is bordered by a small but typical salt marsh. Unfortunately, in 1991 Hurricane Bob washed away a few of the northernmost cottages on the Point. On the Outer Cape, houses fall into the sea when the land beneath erodes during unusually stormy and high tides. On Buzzards Bay, houses are washed away by the storm surge in hurricanes. The land on which they sit often remains pretty much unchanged after the storm.

In 1995 the destroyed cottages were replaced by five new ones on tall concrete stilts. Several images come to mind while viewing these cottages, including those of a lunar landing module and a "Texas Tower" type off-shore drilling platform.

Cottages built on stilts are commonplace on Cape Cod. The purpose of all is the same, to provide enough space below for the storm surge to pass underneath a well anchored home.

The five Bourne houses are the modern high-tech variety, bearing little resemblance to others and giving the impression that the deliberate intent is to proclaim boldly their difference. The lots are small, and the legs so high, that the resulting appearance is extremely tall and awkward. The legs are of square steel and concrete construction, as opposed to the more common round wooden ones. The small houses above, one being a somewhat scrunched two-story saltbox probably cramped by the need to get the entire edifice under a height limitation, are completely out of proportion.

Not too far away stand two similar, but less incongruous, new cottages. The closest, just across the marsh, seems larger, but has lattice work disguising the missing foundation. The other also sits on open concrete stilts, which are shorter and more widely spread, with a larger, one-story house on top. In both cases, the resulting proportions are much more attractive.

The five stilt houses present a very blunt statement that technology can now overcome the natural landscape on Cape Cod. In so doing they represent a stark contrast to one of the most noted features of Cape Cod history and architecture, the "Cape Cod" house, which was traditionally built and sited to conform to the landscape in such a way as to minimize the unfavorable

---

WALKING BOURNE

Most of the shoreline in Bourne is privately owned and difficult to walk. Circuit Drive is an attractive walk through a picturesque area.

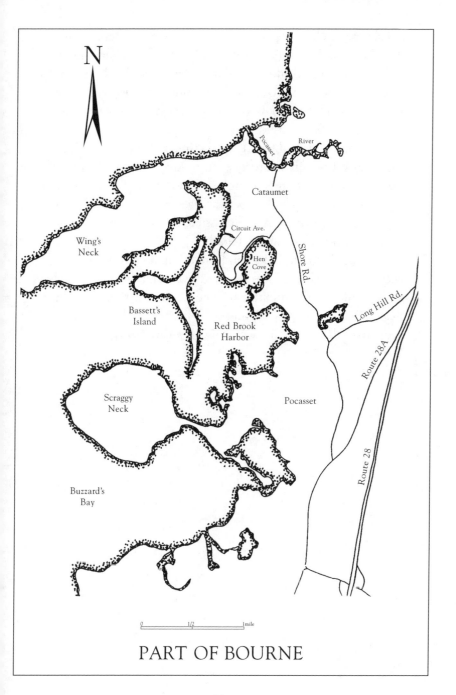

PART OF BOURNE

impact of the elements, rather than to overcome them. I am told that these stilt houses qualified for normal septic systems, but new technology—self contained units—can be utilized in similar places elsewhere. As technology continues to progress, we may see more such homes, not only on our beaches, but out to sea and on our inland wetlands as well.

It is much more difficult to be critical of the owners than of the houses. They undoubtedly came to love the sites and took the only avenue open to rebuilding, codes now permit reconstruction on such sites only if high water can pass beneath. Indeed, as long as a form of construction is permissible on such scenic lots, many owners cannot afford not to rebuild.

Maybe I am a hidebound traditionalist who cannot get accustomed to change. In a hundred years if all the houses are like these, it may be the traditional Cape Cod house that looks out of place. But, to me, on Patuisset Point, the "old" is far more attractive than the "new," as is the case with so much of Cape Cod.

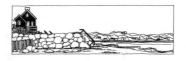

Black Beach, Falmouth
Controlling Development

The Black Beach area of Falmouth includes the largest salt marsh on the Buzzard's Bay shore. The beach itself is a continuously moving, sandy spit created by the process of erosion. Toward the "point," a number of cottages have been built right out on the sand. The last cottage sits on a solid mound of concrete and rock which has probably prevented its destruction. The point does not extend outward into Buzzards Bay, but extends along the shore, ending at the mouth of Sippewissett Creek, which meanders diagonally behind the spit, into the Great Sippewissett Marsh.

The marsh is typical of Cape Cod salt marshes: a lot of rotting detritus and grass flooded at very high tides, meandering streams, and a few bodies of water big enough to be labeled "ponds." Indeed, Black Beach and Sippewissett Marsh would be more in place in Cape Cod Bay, where they have many more siblings, than on the primarily rocky shores of Buzzards Bay.

Great Sippewissett Marsh and the other marshes of Buzzards Bay are different from many marshes elsewhere on the Cape in regard to what man has already done to them. An abandoned earthen railroad trestle slices through the marsh, cutting off the back portion. Route 28A runs along the inland side, providing brief glimpses of the marsh. It is fortunate that this nation built its railroads and its highways when it did. We certainly would not be able to build any now. Railroad tracks in particular have to go straight. How many places go straight for very far without coming to a wetland or a piping plover?

In 1995 Black Beach and Great Sippewissett Marsh became a part of Cape Cod history. Together they became the first area nominated as a "District of Critical Planning Concern" (DCPC) under the Cape Cod Commission Act, which provides a framework for special regulatory powers over such places. As such the area came to play a pioneering role in the struggle over control of development on Cape Cod's shores.

The fifth amendment to the U. S. Constitution, part of the Bill of Rights, provides that private property may not "be taken for public use, without just compensation." Ever since its enactment, those who would like land owners to be able to do whatever they please with their property have seesawed back and forth legally with others who believe it neither appropriate nor desirable to build much in an area such as Black Beach and Sippewissett Marsh. The stakes in such battles, and the intensity of both points of view, get higher as the desirability of a shore-front home and the value of the land escalate upward.

The Cape Cod Commission Act and the county "ordinance" for Black Beach/Great Sippewissett Marsh specify the usual purposes for controlling development such as habitat preservation for rare species (including the threatened but seemingly omnipresent piping plover), drainage, and septic control. If Cape Cod disappears in 10,000 years as geologists predict, piping plovers and lighthouses are sure to be the last two resident species. But, it is easy to discern the real issue behind such designation—the need to determine how much development should be permitted. The ordinance makes repeated statements such as: "The presence of substantial areas of sensitive ecological conditions… render (the) area unsuitable for development."

Indeed, only two qualities distinguish this beach and marsh from many others on the Cape. While the area was not characterized by much development until recently, more seems likely in the immediate future. And, in Falmouth local officials were willing to "test the waters" by stepping into the gulf that lies between preservationists and property owners. Cape Cod has become an intense battleground as these two groups have attempted to shape constitutional interpretation to match their own desires. When does stopping a property owner from building in a fragile area constitute a public taking?

Like everything else in America, the quest to answer this question has become more partisan, political, and divisive as the years go by. As elected officials—and the judges they appoint—change, so does the answer. Court decisions, even at the U. S. Supreme Court level, are constantly changing when it comes to defining under what circumstances zoning becomes a public taking, giving the impression that the process can be as much political as legal. What factors influenced the appointment of the newest judges?

Fortunately, some definitional contours have been added over the years.

Clearly, property owners cannot do whatever they please. Some protections are necessary to shield the public from, for instance, a nuclear waste disposal site next door. It is now commonplace through zoning to exert such controls over development as lot size, setback requirements, and type of use. Not all property owners are entitled to the "highest and best" use, what they must be allowed is "reasonable" use. Reasonable zoning and regulations may result in a decline in property values without constituting a taking. But what is reasonable use?

On the other hand, clearly there must be a public interest justification for zoning which limits the value of a property. But, what constitutes justification?

Two recent decisions have been encouraging to preservationists. One permitted three-acre zoning on Martha's Vineyard and the other prevented a Falmouth property owner from building on fragile, shore-front property not far from Black Beach.

Where does the District of Critical Planning Concern fit in?

Clearly DCPC designation does not supersede the U.S. Constitution, answer the ambiguities of constitutional interpretation, or permanently end the political ebb and flow of regulators and judges.

However, such a designation may well provide more definition and a vehicle for defining "reasonable" public purposes for zoning limitations on development. By so doing, it may make the zoning process less vulnerable to challenge. Falmouth deserves credit for stepping up and nominating the first area. Many other portions of the Cape Cod shoreline require equal consideration. Other towns should consider similar areas for nomination, sooner rather than later. The best way to stop a bulldozer is to prevent it from coming, not to wait until it has arrived before lying down in front of it.

Falmouth seems more willing than other Cape towns to take on such issues. In other places, including Chatham, selectmen have backed off from such disputes rather than expose the town to the possible legal expense of defending against such controversies.

Even in Falmouth, however, the road has not been entirely smooth.

In late 1996 and again in 1997 Falmouth Town Meeting rejected proposed zoning changes in the DCPC. The end result of the designation was thus limited to more extensive Conservation Commission wetlands regulations.

The Falmouth experience thus represents another paradigm of the political complexity of controlling development. In the abstract, voters are almost always for such control. But political scenarios can change rapidly when details make it clear who is impacted and how, and especially when many of those who are affected become active in opposition. Self-interest is often an overwhelming mismatch for public good in the American political and regulatory processes.

Woods Hole
Science and Economics

The village of Woods Hole in Falmouth looks different from any other place on the Cape Cod shoreline. Large ocean-going ships berth at the docks of the Woods Hole Oceanographic Institute, the National Marine Fisheries Service, the U. S. Coast Guard, and the Steamship Authority, serving notice that, by Cape Cod standards, Woods Hole is still a major port. Three-story office and research buildings larger than those elsewhere on Cape Cod, line many of the commercial streets of the village near the waterfront. All of the waterfront, including the inlet of Eel Pond, is lined by seawalls that provide a more permanent and stable shoreline.

Although Woods Hole exudes the charm of the past, everything about it looks prosperous and up-to-date. Its piers and docks—often made of concrete and metal rather than the weathered and beaten wood found elsewhere on Cape Cod—are well maintained and appear fully serviceable.

Indeed, Woods Hole more than looks different. It is different. The village is economically prosperous, unquestionably the most economically productive segment of Cape Cod's shore front. The office and laboratory buildings are home to good jobs—many of them year-round—and the ships sail the world's seas engaged in productive research and other roles.

Indeed, whenever talk turns to improving Cape Cod's economy, sooner or later someone states: "What the Cape needs is another Woods Hole." Cape Codders are not alone. Japan, and Santa Cruz, California and other places, have sent delegations to Woods Hole in recent years to study and duplicate it, right down to its architectural charm. Not only are both of these places far better organized and funded than Cape Cod in their efforts to do so, they have another very real advantage—they understand and appreciate the uniqueness of what they are trying to duplicate—a scientific community with a sense of history and purpose.

Cape Cod does not understand Woods Hole.

The scientific community in Woods Hole got its start in 1870 when the federal government asked Spencer Baird to go to the northeast coast of the country and study the fisheries. Baird could have gone many places, but he selected Woods Hole. It had a railroad, and the marine life was particularly abundant around Cape Cod, where the cold northern flows meet the warmer waters from the south. The three natural corners of Cape Cod geography—Woods Hole, Chatham, and Provincetown—are each situated where differing bodies of water and weather systems meet. (Man manufactured the fourth such corner—the Cape Cod Canal.) In addition, the rocky shores of Woods Hole provided natural harbors for research vessels requiring less dredging than elsewhere on Cape Cod.

Baird began a process that has since spawned many organizations and created a gathering place for literally thousands of marine scientists anxious to compare notes and build upon one another's work.

The first spin-off from Baird's government installation was the Marine Biological Laboratory (MBL) created in 1888 when some of the scientists employed by the Fisheries Service wanted to go to work for themselves on unique projects not directly related to fisheries. The Woods Hole Oceanographic Institution (WHOI) followed in 1930. Then another government agency, the U.S. Geological Survey took its place in the village, followed by the Woods Hole Research Center in 1984 and the Sea Education Association. Today, the Woods Hole Oceanographic Institution remains "the largest independent oceanographic research institution in the nation."

Of course, the larger research organizations continue to give birth to many more small private companies.

The economic impact of Woods Hole on Cape Cod is certainly substantial. WHOI provides 950 year-round jobs. The Marine Biological Laboratory work force is 200 year-round—1,000 in summer.

Like many special places on Cape Cod, the village of Woods Hole has no legally defined boundaries. I once asked a room full of Woods Holers, what its geographic limits are. After discussion, they rejected the school district as a definition, then decided that "Woods Hole is where you get your mail." I did not ask whether the village would disappear if the post office closed. It would probably be too sore a subject after the announced closing of the only bank branch in Woods Hole.

Woods Hole does have a distinct sense of place, however, one that has been shaped by scientists and intellectuals—somewhat unique commodities on Cape Cod. The community fought successfully for several years to keep out a McDonalds restaurant, as well as other similar companies, which residents thought would unduly alter the character of the village. But, even the village's sense of place has its contradictions. When Bank of Boston, the recent acquiree of a Woods Hole bank branch that was once part of Falmouth National Bank, announced the closing of that branch, the news was protested almost as vigorously as the coming of McDonalds. Would it have been different if McDonalds was already in Woods Hole and Bank of Boston was trying move in?

Could the Cape attract "another Woods Hole"—or even expect much expansion from the one we have? Probably not, Woods Hole itself is threatened by a variety of forces:

• Competition threatens from many other places in the world

which want "Woods Hole"—the benefits of a vibrant scientific community—considerably more than Cape Cod does, and are willing to expend far more resources to attract such facilities. Jobs at government facilities have already been going the other way— to competing off-Cape facilities—for many years. Although the Marine Biological Laboratories was able to build a major new building as recently as 1991, the ability to get new projects approved is at best uncertain, as zoning tightens everywhere on Cape Cod, including Woods Hole.

- Woods Hole is also threatened by diminishing federal funding for scientific research. Eighty percent of WHOI's budget comes from the federal government.

- Woods Hole is threatened by too many tourists—the ones who clog its streets and overwhelm its parking lots with cars while waiting for the summer ferries and who come on rainy days to view the aquarium and other attractions of the community itself. On Cape Cod, anything which creates traffic is increasingly threatening, and therefore threatened. The scientific installations in Woods Hole cause some of the village's traffic and parking problems, but they also get blamed for problems which they do not create.

Cape Codders would be wise to understand the economic implications of Woods Hole. Jobs are created where people want to live. Computers have recently made more of the world's good jobs portable than have ever been so in the past. The Cape has perhaps never been in a better position to attract good jobs. Unfortunately, we do not seem to realize it. Far too often, those who represent the business community on Cape Cod give mere lip-service to such phrases as "meaningful year-round jobs," and inevitably revert to discussing how to attract more tourists or how to keep regulations from impeding land development. Cape Cod does not need any more restaurants or gift shops. Yet, there will probably always be an army of romantics desiring to start one or the other, and an even bigger army desiring to help them, particularly if it means selling them some real estate.

Cape Cod should take a close look at Woods Hole. There may never be more scientists on the Cape, but there is still much to learn in Woods Hole, for both sides of the ongoing Cape Cod conflict between creating job opportunities and controlling development. Of course, Woods Hole, like the other charming villages of our past, would not be there without the unique (to Cape Cod) facilities necessary to serve it—facilities which for the most part probably could no longer be built in today's environment. Cape Cod will never attract more meaningful jobs unless those who covet them take their eyes off tourists and real estate development, and start aiming their activities in the direction of meaningful job creation.

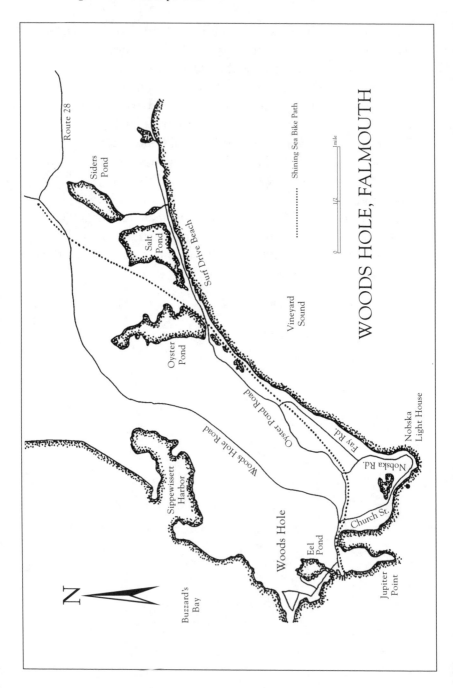

WALKING IN WOODS HOLE AND FALMOUTH

The Village of Woods Hole is scenic and historic. The docks and buildings of the many marine-based institutions make a tour of the village extremely interesting.

Nobska Point, reached from Woods Hole along scenic Church Street, represents one of the most attractive of Cape Cod vistas. Nobska Light, is visible as the walker approaches from either Woods Hole or Falmouth. Martha's Vineyard provides a backdrop for the waters of Nantucket Sound, which in the summer bustle with boating activity of all varieties.

There are a few walking options from Woods Hole or Nobska Point to Falmouth. The Shining Sea Bikeway, following an old railroad bed, passes through woods and along both ponds and the shore. Nobska, Fay, Oyster Pond, and Surf Drive Roads provide an excellent sense of the character of Woods Hole and Falmouth. The shore front is privately owned at the Nobska Light end, hence walking is subject to the limitations of the Colonial Ordinances, although the only sign I saw pointing this out was faded and difficult to read. Surf Drive Beach at the Falmouth end is a typically beautiful Cape Cod Beach. A rocky section of shoreline in between gives the walker a sense of the character of much of the shore at this end of the Cape.

Falmouth Inner Harbor
Swimming Nantucket Sound

During one of the first swims in our trek around Cape Cod Sue developed a sore hip while swimming after a long day of walking. Always concerned about margin for error, I decided to purchase life preservers. I asked the salesman at Goose Hummock in Orleans about the difference between several models.

"Are any better at keeping one alive?" I asked.

"All have the same rating."

"How much benefit will a preserver provide if the going gets tough?"

"Well, in a really dicey situation all one will do is help to find the body."

Thus reassured, I bought two. They helped immeasurably, facilitating much more relaxing swimming—and resting.

Much to Sue's delight, as we reached Nantucket Sound it seemed as if we were swimming more than we were walking. But, a number of challenges were present to some degree in almost every swim, including fatigue, boats, current, appropriate route selection, and as fall progressed—worry about water temperature.

Boats were unquestionably the most dangerous feature of swimming channels. Scouting the channel from the side, it was easy to ascertain that if boat operators did not sight a swimmer over the bow a long distance ahead, they were not likely to see one up close. And, they were certainly not going to hear one.

Whenever possible, I scheduled channel swims so that there would be as few boats as possible, delaying the crossing of many of the busiest channels until after Labor Day.

Our closest encounter with a boat was on a sunny October Sunday at the Falmouth Inner Harbor Channel. As Sue tells the story:

> Swimming channels was always my favorite part of the walks. They represented a little bit more of a challenge and a touch of the unknown, and thereby added to the excitement.
>
> At the end of one portion of the Falmouth walk, we came to the Falmouth Inner Harbor channel where we had left one car. Elliott was going to walk on, and I would drive to pick him up at his finish. Therefore, I did not have to swim the channel. However, I didn't want to miss out on any possible adventures. I did not see any "no swimming" signs and went to the edge. There were some people on the jetty beginning to take an interest. I waited until the coast was clear of boat traffic and slipped into the channel. I had agreed to turn around in the middle and swim back, but I couldn't resist doing the whole thing and kept going. Elliott waited until I was halfway across to start.
>
> Once across, we faced the problem of my getting back. The boat traffic had picked up again. After several vessels had gone by I saw only one boat approaching. As it was a good distance away and going quite slowly, I assumed that I could safely swim. I was halfway across when he tooted his horn. I thought he was being friendly and almost waved at him. I reached the shore and was starting to leave the water when the man, still a good distance away, started screaming at me. Then I heard him bellowed into his radio, "There's someone swimming in the channel." At that point I made tracks—a couple

watching us were ready to greet me as if I had swum the English Channel—but I didn't think it was advisable to wait and see if the radio message had been received at headquarters.

Timing of the tides was also important in channel swimming. The narrowest portion of the channel was always the deepest and had the fastest current. I quickly realized that if I misjudged and got caught in one of these currents, there was little point in fighting it. Life preservers would provide lasting power to enjoy the ride until reaching the "delta" at either end where speed inevitably slowed, depositing sand and whatever else had come along for the ride. These "deltas" often provided the option of wading a longer, half-moon shaped route at the mouth of the channel, rather than swimming the deeper part. Nevertheless, as much as possible, I timed walks to arrive at major channels as close to slack low tide as possible or slightly thereafter, when the ride would be in, not out, if a mistake were made.

This worked well until Popponesset Bay between Mashpee and Cotuit. Sue and I had begun the day by swimming the channels from Menauhant to Washburn Island and on to South Cape Beach in Mashpee. Sue stopped walking there, while I continued on. Because there were swims at both ends of the walk, I had timed low tide for halfway in between and arrived at the Popponesset channel well after low tide. A strong current was flowing in, but I had incorrectly concluded that the volume of flow into this particular bay would not be large enough to be of major concern. Although I knew that the current would be less and the water perhaps shallower on the delta route at the mouth, I nevertheless headed right over to the narrowest part and began to swim.

Ordinarily, I realized that I was in a current if I looked up every few seconds and saw a different place on the far side. This time whenever I looked up the bank was going by as if I were on a speeding train. The ride into the bay was swift, but stopped as quickly as it had started, once I cleared the end of the channel. The swim to the far shore was then easy, but my destination was further into the bay than originally anticipated. After reaching land, I continued the walk along a seawall into Cotuit.

The most intimidating swim of all was yet to come—at Bass River. Again, I was walking alone. Without Sue to charge right in, I felt like the cowardly lion. I thought I knew the river well from previous participation in an annual canoe race, but it was far wider jetty to jetty than I recalled. A steady stream of boats headed out to Nantucket Sound. The wind was blowing out, creating waves which made it look as if the water was going with them. Although I was quite sure that I had timed slack tide correctly, I worried about the current taking me out to sea.

I stood on the jetty for twenty minutes or so, timing the boats and waiting for a gap. But, I had another reason for pausing: I simply hoped the whole prospect would go away. Perhaps Huck Finn would come by on a raft and offer me a ride across.

Even when it comes to procrastination, enough is enough. An appropriate interval in the spacing of the boats finally developed, and I knew that there would not be a better one. There was no current, just waves. Several boats crossed my bow, but none threatened as I crossed their path. The swim itself went very well. But as I staggered up the far jetty, I was surprised to almost collapse from exhaustion. "Pumped up" in my intense desire to get to the other side, I had unknowingly swum far faster than I realized.

By fall, we began to worry about another hazard, the question of whether or not the water would get too cold to swim in. Right through our last swim in late October, it never did. If anything, as the air temperature dropped, the water seemed to get warmer.

All told there were 31 swims on our way around the Cape. As Sue put it, "Swims always added zest to the walks, and encouraged me to go another day."

Waquoit Bay
Public Purchase of Land

Access to the shore, whether active or passive, has always been critical to the strength of the Cape Cod economy. Of the 427 miles of Cape Cod shoreline, 283 (66%) are privately owned. Private ownership is not evenly distributed. Only 44% of the coastline is privately owned in the six National Seashore towns, compared to 83% in the remaining nine towns. In Barnstable, the most populated town, only ten of 61 miles are publicly owned—almost all on Sandy Neck, well removed from the most populated areas of the town.

Although the publicly owned portion of the Cape's shoreline has increased in this century, access has not. To the contrary, as the number of private homes increases, so do "no trespassing" signs on land that has long been privately owned, but not previously closed to human passage. This trend in turn puts more pressure on existing public beaches and those portions of the coastline which are available to full or limited public use for activities ranging from aquaculture to swimming and walking, or just plain viewing. Of course, public ownership of land has become an issue of growing importance on Cape Cod as well.

Recognizing the problem, on August 7, 1961, the U. S. Congress passed the Cape Cod National Seashore Act. Before its passage, only five eighths of the shore front currently in the National Seashore was publicly owned. The $16 million appropriated in 1961 represented the first time which the Congress had ever authorized the use of public funds to purchase land for a national park. (The amount ultimately expended was more than doubled the initial appropriation.) Prior to that time every national park created was either on land already owned or land donated to the park service by owners desirous of preserving it.

Many Cape Codders did not support creation of the National Seashore. Polls in Truro and other towns indicated a majority were not in favor. Town leaders of the six towns within the proposed park met to coordinate opposition. The strongest initial support came from the tremendous number of off-Cape people who had come to know and love the Cape.

Eventually the legislative process became a debate primarily over boundaries, the areas in question being the Provincelands in Provincetown, the great ponds in Truro and Wellfleet, Great Island in Wellfleet, the Fort Hill area in Eastham, and Morris Island and Stage Island in Chatham. All but the last were eventually included.

Many of the most outspoken opponents of the National Seashore lived to acknowledge their mistake. It is difficult to envision what the Outer Cape would have become if it had not been created. With over five million visits recorded a year, Cape Cod National Seashore is one of the most utilized of all National Parks.

Since 1983, the primary area of public purchase of land on Cape Cod has been Waquoit Bay between Falmouth and Mashpee, where the federal government, the State of Massachusetts, and the Town of Mashpee have purchased more than 1,650 acres of land for millions of dollars, creating South Cape Beach State Park, the Waquoit Bay National Estuarine Research Reserve, and the Mashpee National Wildlife Refuge in the process. Several miles of valuable shoreline have been preserved.

Combined, the state and town beaches at South Cape Beach State Park include some of the most beautiful sandy beaches on Nantucket Sound, reducing the pressure on other public beaches on the Cape's busy south shore.

The purchase of Washburn Island in the middle of Waquoit Bay, with more than three miles of shore frontage, as part of the National Estuarine Research Reserve represents one of the most substantial reclamations of waterfront property in this century. During World War II this island was the site of enough military activity to necessitate constructing a bridge to the end of the adjacent point in Falmouth. The remains of the paved road on the island up to the bridge are still there, as Sue and I discovered, along with those of a dike which shortened the distance.

Sue and I canoed over to Washburn Island from the town landing on Seapit Road in Falmouth and walked its perimeter on a sunny October Sunday. The island illustrates how fast land returns to its natural state when intensive use by mankind ceases. Although the remains of foundations and roadways are visible in several places, the experience there is largely that of a wilderness. We passed only four people in the several hours of our perambulation. Two, middle aged ladies sunbathing fully dressed in their October clothing, after asking the shortest route back to their boat, gushed out, "I think we have stumbled on a wonderful secret here." Ten campsites dot the island, the only method of visiting for more than a self-propelled day trip. The sites illustrate how even light use can leave its marks, however. The pathway from each campsite down a short bank to the water is eroded where the occupants have scrambled up and down in the same place over and over again. The scarring would not be so noticeable were it not in such sharp contrast to the pristine condition of the rest of the island.

The other major public purchases of land in the area includes 400 acres on the Quashnet River to refurbish a trout stream and over 400 acres in the Bufflehead Bay area of Mashpee.

Waquoit Bay National Estuarine Research Reserve is one of twenty-one designated reserves in the country. Its primary purpose is to undertake and publish research of use to "managers" of other similar bodies of water, not to preserve land or serve as a nature center.

The recently created Mashpee National Wildlife Refuge contains over 5,800 acres within its boundaries. Slightly over half of this land is presently publicly owned and preserved. Since the designation, a small amount of privately owned land within the boundaries has been developed. The remainder threatens to be the scene of a race between developers and preservationists seeking funding.

No management plan has yet been prepared by the federal government for the Refuge.

In hindsight, every public purchase of land on Cape Cod was a wise decision, and every failure to do so, including that of Morris Island in Chatham, was not. Morris Island is already on the inevitable road to conflict between the large private homes which have since been constructed there, and use of the abutting Monomoy National Wildlife Preserve. Inevitably this conflict will only grow, particularly if, as anticipated, the Monomoy Islands reattach themselves to the mainland at which time many people will attempt to reach the Monomoy Refuge by land—via Morris Island.

Why, therefore, do people continue to oppose the public purchase of land through such programs as land banks? Pressures on shore utilization are greater now than they were in 1961, and they will clearly be much greater in the future.

There would appear to be three reasons for such opposition:

• First, land development is the major industry on Cape Cod, an industry like many others, scared by and resistant to change.

• Second, opponents of the National Seashore feared that public purchase of so much land would undermine the tax base. This did not happen. Taxes are lower in the National Seashore towns, and it is well documented on Cape Cod and elsewhere that rapid development increases local taxes, while the public purchase of land does not. Still, opponents of proposals such as land banks continue to find it effective to sing the same old song about undermining the tax base.

- Finally, trade associations in the United States are like Roman gladiators—always looking for a dragon to slay so they can justify their dues and continued existence to the membership. National, state and local realtor associations have focused opposition on land bank issues. Realtors on Nantucket and Martha's Vineyard initially opposed similar proposals, but since their passage have come to believe they are in the interest of the real estate community. Barron's, hardly the voice of conservationists, recently labeled the existence of land banks as "one reason" why property values in Nantucket and Martha's Vineyard are faring better than in similar areas without land banks.

But, once we take positions in America, it becomes increasingly difficult to abandon them.

Cape Cod will go on needing to purchase more public land for a long, long time. Special interest groups will go on opposing such purchases for a long, long time.

Walking the Shores of Cape Cod

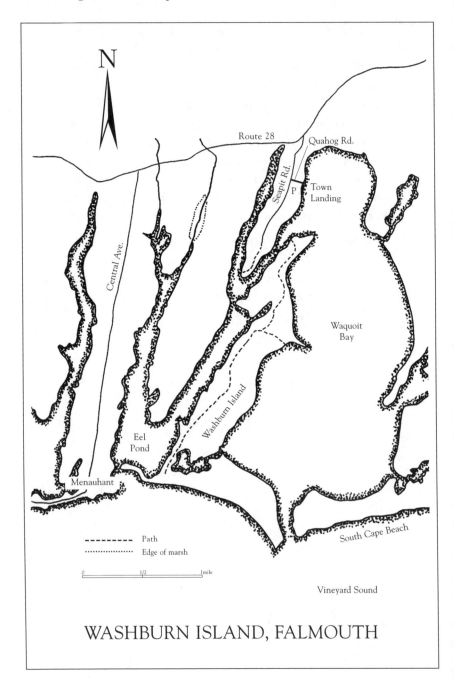

WASHBURN ISLAND, FALMOUTH

WALKING WASHBURN ISLAND IN WAQUOIT BAY

Washburn Island in Waquoit Bay, once reached by bridge from Falmouth, was the scene of considerable military activity during World War II. It is now state owned, but is only accessible by boat, from Seapit Landing in Falmouth or the South Cape Beach area in Mashpee. Next to Monomoy Island in Chatham, this island is the Cape's most wilderness like coastal experience. Remains of a failed groin built to protect the beach and of various World War II vintage structures may be seen along the scenic barrier beach on the Nantucket Sound end of the island. There are trails on the middle of the main portion of the island, but none on the shoreline, which in places contains obstacles including marsh, brooks, and overhanging vegetation.

Osterville
A Large House by the Sea

Until this century, almost all of the residents of Cape Cod lived here year-round. They did not choose to live right on the shore. If they did, their homes were modest. The grandest of pre-twentieth century buildings were the stately homes of sea captains, lined up on village main streets within viewing distance of the water from their widow's walks, but well removed from its stormy fury. The captains saw enough of that at work.

This all began to change at the close of the nineteenth century, when improved transportation brought the Cape within range of summer residents and vacationers. These groups brought two qualities previously lacking. The first was greater wealth than had ever been accumulated on Cape Cod with which to finance homes. (Even the sea captains' houses were "trophy" homes, built to reflect wealth accumulated and time spent elsewhere.) The second was a desire to place their homes right on the water's edge inasmuch as the occupants came primarily to enjoy the water during comparatively benign summer weather, rather than to make a living from it the year-round.

In the twentieth century, many neighborhoods of magnificent waterfront homes have been built, including Scraggy Neck in Bourne, Penzance Point in Woods Hole, New Seabury in Mashpee, Oyster Harbors on Grand Island in Osterville, Great Island in Yarmouth, Old Mill Point in Harwich, Morris Island in

Chatham, and others. The homes therein are large, attractive, and anything but modest—residentially speaking the ultimate in fulfillment of the American dream.

In a free economy, scarce resources are purchased by the highest bidder. Since scenic land became the primary commodity which Cape Cod has to offer in economic trade, almost all of the buyers in such neighborhoods have been from off-Cape. The larger and more expensive the home, the less likely it is to be occupied year-round. A few local realtors and developers who were driving beat up dump trucks in the early 1980's, later bought waterfront homes following the real estate boom of that decade. Many have since had to sell their homes to repurchase their truck. What goes up, usually comes down.

Although such neighborhoods are attractive and well maintained, they have a certain sense of hands-off sterility about them. One fall morning I walked the roads of two "neighborhoods" which share the same peninsula in Nantucket Sound. In one neighborhood, the houses and lots are large and of recent vintage. Indeed, most are far larger than necessary to shelter even a very large family. But, even walking on a public road on a fall day that was sunny enough to send spirits flying, the feeling was anything but welcome. Gates block several driveways along with "beware of dogs" signs. Side streets lead only to larger houses which block access to the water beyond. There were few or no signs of current occupancy in the houses. The atmosphere given to the casual passer-through was best characterized by one "no trespassing" sign at the end of a trail to a duck blind on which several choice profanities had been scrawled.

The settlement at the end of the point is as different as it could be. Built up years ago, small cottages—only a few of which have been improved even though many are now occupied year around—crowd small lots. If the neighborhood were not in such a beautiful place, it would probably be considered a little seedy. But it is, and everyone has access to the water at the end of a series of dirt roads. The occupants of every car which passed, waved, giving the impression that such a neighborhood encourages friendship.

Most of the inhabitants of shore-front neighborhoods have traditionally been from off-Cape, but their composition has changed with the passage of time. According to local historians, as our shores were developed earlier this century, educators were one of the more preponderant groups to summer here. Prices were then affordable, and teachers had the entire summer off to reflect and recharge, making the trip more worthwhile when limited transportation made it far more time-consuming to get here. Now it takes more financial success than is typically available in academia to afford such a location. Ultimately the coast may be inhabited primarily by professional basketball players, the profession which our society currently chooses to reward most richly.

In order to limit access, most such neighborhoods have a guard house on the entry road. The one at Oyster Harbors is said to turn back 10,000 cars a year. The sentinels are probably there for a number of reasons: to insure privacy, particularly if the area is home to any famous residents who might themselves be tourist attractions, to prevent crowding of the beaches, to increase security, and to give an appearance of exclusivity, which sells well in America.

While touring Cape Cod's perimeter, I did not approach such neighborhoods along their access roads. For the most part I walked in and out along the shoreline. For example, I viewed Oyster Harbors from Dead Neck and Sampson's Island—two connected sand spits providing close range viewing of the south or coastal end of Grand Island across the "Seapuit River," a narrow saltwater channel. I swam to these spits from Cotuit, and from them to Wianno. Later, I cruised around all of the island on a tour sponsored by the Barnstable Conservation Trust.

Sometimes I had advance permission before entering such neighborhoods, and sometimes—primarily in the off-season—I did not. Only once was there a guard policing access along the beach. I passed by with a friend from the neighborhood. One of the more surprising passages was the ease of walking the beach in front of the Kennedy Compound on Squaw Island in Hyannisport.

As they multiply, such houses and neighborhoods clearly change the character of Cape Cod, in some ways for the better, but in some ways not.

The residents are good conservationists. Typically there are fewer houses per acre. And, the more guarded the enclave and the fewer the visitors, the more pristine the beach. The beach on the outer side of Great Island in West Yarmouth is as clean as any I saw. In contrast, our national parks are threatened by over-visitation.

Such neighborhoods are clearly good for the tax base; they pay more in taxes than they need to receive in services. Far more often than not, the residents are good people to have in town, contributing much time, money, and expertise.

However, the larger the portion of the Cape's shoreline which is bought and closed off by a small number of people, the more that everyone else—and the Cape Cod economy with them—get trapped inland. The shoreline is all there is to the Cape Cod economy. Without it, the Cape is not much more than a pile of sand, almost totally lacking in economically meaningful resources. Every time a segment of shoreline is closed, a piece of the economy is retired. Great Island alone has more private shoreline than there is in all of Brewster, and almost as much as there is in the remainder of Yarmouth.

Everything is relative. While my family was building a new home on a quiet pond in Brewster a number of years ago, I read an article by a woman who had vacationed in Chatham for many years. She wrote that an annual walk through the woods around the pond on which we were building had been one of the highlights of each summer's vacation. I read this with a sense of regret that new homes might diminish her pleasure, along with a warm feeling that someone from Chatham would make an annual pilgrimage to our pond. It occurred to me that it might be nice to build a trail around the pond so that both residents and visitors could enjoy it. I suggested this to two neighbors, both retirees from big cities. They looked at me as if I were crazy—if not dangerous. I let the matter drop.

Although wealthy homeowners are not the only ones to close the beach, the wealthier the purchaser, the more likely one is to expect, seek, and be able to pay for exclusivity. There are, however, tools which could be utilized more vigorously to maintain public access. For example, every time owners build a dock or a revetment across the low tide line, they "trespass" on public property. More could be asked in return.

Cape Cod needs to be concerned about how much of our shoreline is closed to public access. Many public beaches are already too crowded and there is a growing need for other types of public waterfront facilities. As in so many ways, the Cape needs a better plan for our shores. How much of the coast can we afford to privatize, and in what ways, before we have critically diminished the most important engine behind our economy?

I never saw a new house that looked as good as the open space which it replaced. That, in a nutshell, is the Cape's dilemma. Every time another house is built, a seller realizes the value from the land, a buyer is made happy to be the latest to cross the bridge, and a realtor, a banker, a lawyer, and a builder all make a little money. But, another piece of Cape Cod is gone, particularly if the new home is on the shore and the owner chooses to close the beach.

Wianno
Hurricanes

Early this century, my grandparents bought a waterfront lot in Mattapoisett, on the opposite side of Buzzards Bay from Falmouth, and built a small summer cottage. The lot was near the end of a small rocky point which is exposed to quite a bit of "weather" whenever the wind blows from the south, as it often does on Buzzards Bay.

In September of 1938 the cottage was destroyed by the great New England hurricane, the worst this century. Two days later, I was born in New Hampshire.

Considering the 1938 hurricane to have been a fluke, my grandparents and their neighbors soon rebuilt. Initially, there had been a road between my grandparents' cottage and the water, but the shoreline, even though rocky, had been slowly eroding each year. By the time the second cottage was built following the hurricane, the road had been entirely washed away. Rid of this public nuisance, my grandparents and their neighbors built a small seawall in front of their properties. It was made of concrete and rock, about five feet high and only twelve inches thick. Normal high tides stopped at the bottom of the wall. The steps of the new cottage were about 30 feet back from the wall and three more feet higher up—eight feet in all above high tide.

In late August of 1954, when I was sixteen, my family was at the cottage as Hurricane Carol approached. In those days hurricane forecasting was not what it is today. Although radio news the day before told of a storm in the ocean to the south, it did not indicate that a hurricane might arrive the next morning. But, soon after we awoke the ocean began to come over the wall, a sight which I had never witnessed before.

Shortly thereafter, water still rising, my family retreated to another cottage further back from the water. Much of the neighborhood was already gathered there.

Still later, water still rising, we went back to retrieve the family dog, which we had left behind.

A while after that, water still rising, we went back to get the car out of the garage. Too late! The water in the garage was knee deep. This time we had to dodge a considerable amount of windblown debris as we moved from cottage to cottage.

The next move, water still rising, came when everyone vacated the second cottage and walked a half mile up the road to a sizable clearing in the woods which the adults considered a safe place to wait out the hurricane. We did so without memorable occurrence.

When we returned in mid afternoon, every property had been damaged. Our cottage and two others had been destroyed. The boats of the harbor were in total disarray—either smashed on the beach or floating upside down many feet from their original anchorage. The neighborhood swimming raft was wedged beneath another cottage.

Water did the damage, not wind. Our cottage was probably destroyed when another floated into it. The first floor was totally obliterated, but most of the second floor rested about two hundred yards inland, bone dry, having floated there like a house boat. My bedroom was totally undisturbed. The beds were as we left them, as were my two brother's clothes in the dressers and closets. Unfortunately, always prepared, I had quickly packed most of my belongings and taken them downstairs. They were never found.

The road to the point was cut off from civilization for another day while bulldozers removed the debris from another enclave of cottages which had been totally demolished. My family spent the next night in the slightly damaged cottage of friends, then returned home in a car borrowed from relatives in Boston.

This time it took six years to make the decision to rebuild. By then there was a joke in the neighborhood that the most reliable method of predicting bad hurricanes was to count the number of cottages on a nearby sand spit. Whenever most had been rebuilt, it was time for another storm. Presently the spit is occupied by thirteen stilt houses similar to those I encountered in Bourne.

The third cottage which my grandparents built was again seasonal, far more modest than its predecessor. It, too, was constructed one floor above ground level on steel columns which allow the storm surge to pass beneath. Such construction had withstood Hurricane Carol far better than conventional techniques. It is still there, as is the seawall. As a student, I was frequently there on vacation in the late summer hurricane seasons of the sixties. Several times I left the cottage to find a safer vantage point from which to watch.

Although the seawall is still there, a gap left in the middle at the end of a public way has created an ever-growing breach which creeps around both ends, threatening the future of the entire wall, which has been supplemented by loose rocks piled against it. At high tide, the water now rises about a foot above the bottom of the seawall—the result of sea level rise.

My grandparents and their neighbors also unknowingly pushed the large rocks onto each side to clear a number of small "beaches" between the rocky segments. As a result, each year the level of the beaches gets slowly lower, undermining the wall, while the level of the rocky sections rises in between as the rocks catch sand. The wall may not hold the sea back forever, but the lot would have been gone long ago if it had never been there.

We know considerably more about hurricanes now than we did in 1954. They are circular storms which move counter-clockwise, hence, although they always come to New England from the south, the wind can be blowing in any direction depending on which part of the storm is passing overhead. As a result, the impact of a hurricane on relatively close together areas can be very different. For example, if a hurricane with 75 mile-an-hour winds is moving north at a speed of 20 miles-an-hour, the land which the eye crosses will first experience a 75 mile-an-hour wind from the east, then a 75 mile-an-hour wind from the west. To the west a 55 mile-an-hour wind blows from the north, but to the east a 95 mile-an-hour wind comes from the south. Meanwhile, a mound of water several feet high moves along in the low pressure of the eye. On Cape Cod, the strong south winds on the east side push a considerable storm surge into the bays and inlets on the south side. But, in contrast, little or no storm surge is experienced on the west side of the hurricane, where the lesser winds from the north come off the land.

Although exceptions exist, including the 1938 hurricane, the area of greatest damage is often quite limited and almost always to the east of the eye. The water does the damage! The west side, with lesser winds, is characterized primarily by more rain, as the diminishing winds reduce the storm's moisture carrying capacity. Hurricanes are rated in severity from category one (74-95 MPH winds) to category five (winds over 155 MPH). Because hurricanes require a water temperature above 80 degrees Fahrenheit to maintain full strength, and the water temperature immediately south of New England seldom reaches that level, most hurricanes dissipate in strength as they reach New England. As a result, the area experiences hurricanes in the upper categories with less frequency than do the southern states.

The three hurricanes of most historic note to southeastern Massachusetts and Cape Cod were the 1938, 1944 and 1954 hurricanes. Several remarkable similarities characterize these storms. The eye of each passed over eastern Long Island—the worst place the eye of a hurricane can go for Cape Cod is a few miles west of Long Island. Later in each of the three years a hurricane eye went over the Cape itself. No one remembers these storms; the damaging portion of them was out to sea. But if the eye ever crosses the coast between New London, Connecticut, and Providence, Rhode Island, watch out on Cape Cod!

All three of these fierce hurricanes had been traveling in a straight line for a long distance before they hit, maximizing the distance of the fetch, and therefore the size of the waves on the east side of the storm when it hit. A hurricane traveling on an erratic zigzag course has a lesser fetch.

Finally, all three of the hurricanes arrived in New England traveling at a speed of 60 to 80 miles-an-hour, having accelerated from Cape Hatteras, North Carolina. This was significant in two respects. Inasmuch as no one anticipated this acceleration, the hurricanes were not expected to arrive in New England as soon as they did. And, even a moderate hurricane with dying winds becomes quite strong on the east side when the storm is traveling 80 miles-an-hour. Hence top winds of well over 100 miles-an-hour were recorded in each of these storms. The peak recorded wind was 155 miles-an-hour (at the Blue Hills Observatory in Boston) during the 1938 hurricane, even though it was not a category five hurricane. If a hurricane is accelerating as it passes North Carolina and the eye is headed for landfall slightly west of Cape Cod, the time has definitely come to evacuate the Cape's south coast neighborhoods.

Old timers remember the 1944 hurricane as the worst ever to hit Cape Cod. Those in 1938 and 1954 did their worst damage to the west, roughly between Narragansett Bay in Rhode Island and Buzzards Bay. But, the 1944 storm was moving on more of a northeast course than the other two as it crossed Long Island, bringing it closer to the Cape. It is the only major storm where the worst damage was concentrated in the Cape's most populated area, Hyannis and the other villages of Barnstable.

The storm surge which hit Wianno in the Osterville section of Barnstable brought water measured 10.1 feet above normal high tide. It eroded 40 feet off of the Wianno bluff from Seaview

Avenue to the Osterville Cut. At the end of the storm, water was washing against the foundation of the Wianno Club, which observers say would have been lost if five more feet of the bluff had been lost.

Just to the east, according the Cape Cod Standard Times, cottages at Long Beach in Centerville were "tossed around as a boy would kick a football." The September 21, 1944 edition of the Barnstable Patriot tells stories such as those of Mr. William White and Miss Frances Jennings. Mr. White, custodian of the Oyster Harbors Riding Stable, waited the hurricane out on the roof of the stable, after having released six horses so that they could swim to safety. Meanwhile, Miss Jennings spent an hour in the water with her father, at night, not knowing where they were. The pair crossed East Bay with their cottage floating nearby after it had been torn loose from Dowse's Point. After finally reaching the other side and finding the cottage's kitchen to be bone dry, Miss Jennings prepared tea.

Troops and German POW's from nearby Camp Edwards helped clean up after the storm. One Osterville old timer remembers being without power for 27 days.

Following this hurricane, Herman A. MacDonald, the state's Commissioner of Public Works, suggested building a seawall around Cape Cod to reduce the damage from future storms.

The expensive homes at Wianno have been protected since by regular groins (rock piles constructed perpendicular to the shore to trap lateral sand flows) between each property and seawalls, revetments, or rip rap extending eight feet or so up the bank. No more ground has been lost, but water reaches the walls in some places even at half tide. Although prior to 1944, the shoreline at Wianno was open; now an uninvited walker can only proceed by passing "no trespassing" signs, making it difficult to walk this beach except in the off-season when most of the houses are unoccupied.

Forecasters believed, until El Nino in the summer of 1997, that we had entered a period that will be marked by more hurri-

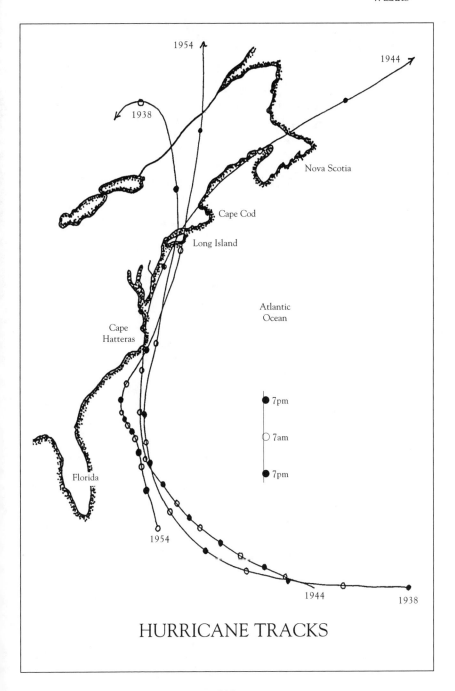

HURRICANE TRACKS

canes. The next hurricane of the magnitude of that of the 1944 debacle will change the Cape forever. Most current residents, numbed by years of false alarms, have no concept of the damage wreaked by such a storm. The initial impact will be disastrous. But, over time, the result may be beneficial if it leads to a greater understanding of what can sensibly be done on our shores and what should not.

Craigville Beach
Public Beaches

With the exception of public beaches, most of the Nantucket Sound shore is privately owned. As a result, many visitors only see the shore at public beaches.

Craigville Beach in Centerville is known as a young people's beach. In the words of the lifeguards, the boys strut their stuff up and down the beach, while the well-oiled girls sun themselves in their "pocket bathing suits," turning their towels like sundials as the day progresses.

To me, beaches are for walking and exploring. Swimming is a way of getting from here to there. Cape Cod has 97 public beaches, some miles long, and many more private or unnamed ones. For someone willing to walk and explore, it is not hard to find privacy on the beach, even on the busiest summer day. On a sunny summer day thousands of people visit these beaches, as we all find out on a rainy day when they go shopping instead— or try to until they get stuck in traffic. There were 4,931,330 total visits to the National Seashore beaches in 1996, led by Herring Cove Beach in Provincetown with 749,426. Given the length of the summer season, that comes to over 10,000 visitors to that one beach alone on a good summer day. Craigville and companion Covell's Beach have 747 parking spaces managed by the town, and many others managed by private operators. The crowds are estimated in excess of 4,000 on a summer day. The revenue produced for the town and local business people is significant. Beaches are the bedrock draw of the Cape Cod economy.

As I walked through the public beaches, I observed that most of the occupants are either sitting and reading or lying and tanning. Why do people come to the beach, I wondered, and even more, why do they lie down? They don't go to the gravel pit to lie down, they would not dress like that anywhere else, and they all say that they don't like crowds—but beaches are where the crowds are on Cape Cod in the summer.

In the 19th century, people went to the beach for their health. The fresh, salty air was thought by doctors to be even better than mountain air: relaxing, an aid to sleep, the appetite and digestion, as well as invigorating. Saltwater was believed to stimulate the skin and circulation, and exposure to the sun was believed to supplement the medicinal powers of both the air and the water. Now, we know that prolonged exposure to the sun causes skin cancer, and that other forms of cancer have elevated rates on Cape Cod, perhaps because all of the air in the country—industrial pollution included—passes over Cape Cod.

But, people come to the Cape's beaches in ever-growing numbers, all kinds of people. Girls come to sun their bodies and get tans, boys come—until they are married—because that is where the girls are; and families come because it is a relaxing place to watch the kids. Others come to experience nature in the multiplicity of forms on display.

Although there are exceptions to everything, it is my impression that females like beaches better than males. They like to sun their bodies, and being more patient, they know how to relax better.

Women—particularly young girls—like a tan. When asked, almost all say it looks good: "makes me look ten years younger" or "ten pounds smaller," "brings out my freckles," or "I stop just before my freckles come out." One mother told me that "a tan worked for me when I was younger" (she still cultivates one), and described her bikini clad daughter as "a flower trying to attract a honey bee."

There is more to the sun than just a tan. Females also like to warm their bodies. Many report feeling energized by the sun.

Men are more skeptical about all this. As one summed it up, "the media has gotten us all to believe that tans look healthier. They are just a placebo for greater self-image."

Of course, selecting the right bathing suit is very important to this process.

Every year the suits seem to get skimpier. Mary Pipher, one of the nation's preeminent counselors of young girls, stated in her book, Reviving Ophelia, that she never met a young girl who liked her body. You would never know that at the beach. "Thong" suits, not much more than two ribbons around the waist and through the crotch, bare everything including the buttocks. Wags label them "butt floss."

John R. Stillgoe in his book, Alongshore, states that bikinis are a "badge of freedom from fashion dictates, an advertisement that its wearer not only recalls long-past days of sartorial abandonment but remains convinced that urban, middle-class prudery cannot last."

One girl told me that she wears a bikini because she likes the sun, but "there are some places the sun should not go." An older woman said that when younger, "she liked the feeling when naked of being tanned everywhere but in sensitive spots, and thought other people might, too."

Of course, not every woman believes that she can wear a two-piece suit. Reportedly, a 1995 survey for DuPont Corporation, the manufacturer of Lycra, indicated that seven out of ten American women would rather go to the dentist than shop for a bathing suit. Large department stores run classes on how to shop for a bathing suit. New suits promote such added features as "bust maximizer," "hip/thigh minimizer," and "tummy/midriff toner."

For many, going to the beach is a family activity. There is something for everyone to do. Mothers tell me that their kids are always happier: they can splash, explore for critters and shells, or dig and build castles all while under mom's watchful eye while she reads or suns. Those fathers who come along, can choose whether to join the kids and mom, or go off and find some other activity with the guys. Brian Shortsleeve, Editor and Publisher of Cape Cod Life and father of a young son, wrote a column in his July 1997 issue on the intricacies and joys of digging and building. He wrote, "Once you get a good location, you'll know it right away. You can feel it. A slight dig lets in a little water and the occasional waves provide the depth you need, but do no real damage. Now, you're ready to go to work. My favorite tool is a small metal shovel three or four inches in width. But sometimes Josh wants to use the shovel and I have to use the pail. I explain to him that I need the shovel to get the digging of the canal started. But he doesn't listen to reason. What does someone less than three years old know about this sort of thing anyway?"

Everyone in the family is glad to be removed from the everyday stresses of the home and workplace.

In many families, the members all have their own beach rituals. My friend, Gordon Wright, says that going to the beach with his wife, Carol, is like going to church. She takes a good book, walks beyond the crowd, angles the towel to the sun and the pitch of the sand, scrapes the sand for comforts sake, and applies plenty of sun tan lotion. Then, she says, "It is like surgery, I'm only semi-conscious." Gordon takes a chair and an umbrella—he likes shade and fears skin cancer—then faces the water, not the sun. I asked both whether they like sand. Carol replied that it has its place, which is not on the towel or in her book or bathing suit. She added that "Gordon has no beach etiquette." Maybe it is because of such charges that many men do not choose to accompany their wives to the beach, preferring to play golf, go boating, or engage in some other activity.

In contrast, I asked one woman in the neighborhood why she does not go to the beach. She responded that when she got married, her husband had dated "a lot of women" who took him to the beach. He is light skinned and endured a considerable amount of suffering before he learned to wrap himself "like Lawrence of Arabia" in a towel, and then not to go at all. By the time she met him, he was no longer interested in "beach blanket dates."

Beaches have many natural attractions. One cool Memorial Day weekend, opening time for Craigville Beach, I observed a middle-aged woman, clad in a winter parka, wading barefoot along the water's edge. I asked her why she comes to the beach. After initially responding that "the sun rejuvenates us," she continued that there is something much more "esoteric" than that, the beach is where "all of the elements which we are made of are present—fire, water, earth and air." She really must have been in another world, her feet were tinged with blue. Others praise the beauty and vastness of the water, the scent of the salt air, the cool ocean breeze, the roar of the surf and its ability to drown out the more mundane sounds of life, and the pleasant feeling of the sand—"like a Japanese foot massage when it gets between the toes," I was told by one woman walking along the beach who offered to give me such a massage. "Did you know that feet have the third most nerve endings of any place on the body?" I was not told what two places have more.

Not as many people actually swim at beaches as I would have thought. Most of those who do go in quickly just to cool off.

Many people use different beaches for different purposes. One friend has one quiet beach for reading, another for walking and exploring, and goes to Craigville whenever he wants to watch people and listen to their conversations, which he considers highly entertaining. Several others, comparing beaches, indicated that "going to Craigville is like going to New York City."

The Cape has its gay beaches and its nudist beaches. (Usually by the time I hike to one, I am too tired to look.) Although the parking lots at Craigville are locked each night at 9:00 pm, I am told that an entirely different set of romantics descends on some of the National Seashore beaches shortly after that time. I wonder if they are included in the official counts.

There is, of course, one common element among all the reasons why people come to beaches. They provide a sense of relief, whether it be from dress codes, standards of behavior, or other stresses of everyday life.

Girls can be girls, boys can be boys, kids can be kids, and adults can be whatever they want to be. I choose to keep on walking, looking, and thinking about what I see.

WALKING NANTUCKET SOUND IN BARNSTABLE

Most of the Barnstable shoreline is privately owned and very difficult to walk. Craigville Beach and Long Point are the only publicly accessible portions of the map.

Craigville Beach

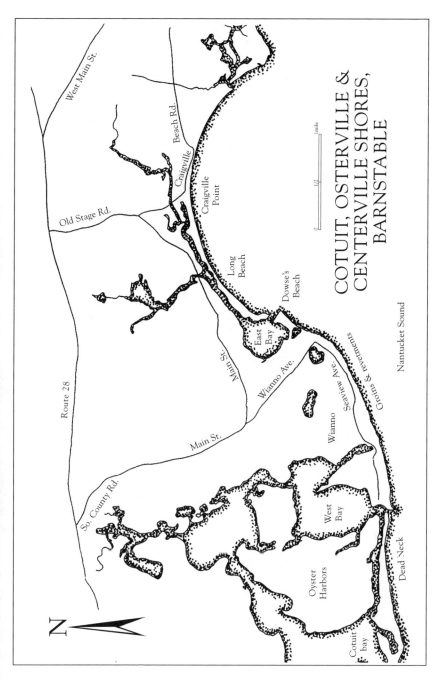

Nantucket Sound
Trespassing

Four security vehicles were present, three with lights flashing, when Sue and I were "removed" from a private shore front.

The "neighborhood" was a large one, controlling a considerable expanse of waterfront. We entered it along the water line at low tide early one Saturday morning as part of my walk around the Cape. As long as our feet were in the water, we were legal, but because of the greater ease of walking, they were out as much as in. We went a long way uninterrupted, although I am quite certain that I saw a guard observing us at least once. We exchanged friendly banter about the possibility of being stopped. Sue did not believe that anyone would question the presence of a friendly middle aged couple.

We approached a young family on the beach. "Wait and see," Sue said, "they will say, 'good morning, isn't it a nice day.'"

"Good morning, isn't it a nice day," said the young father.

"Good morning, yes it is," said Sue, followed shortly thereafter by, "See, I told you so."

"It is not over until the fat lady sings," I responded. "This place undoubtedly has a security force and if anyone is concerned, they will call it, not try to stop us themselves."

In hindsight, I think that my trip around Cape Cod may have reached a phase of civil disobedience. I did not seek or desire to be stopped, but I did want to undergo, indeed to enjoy, all of the experiences which the shores have to offer. Avoiding any place where we might be stopped would have eliminated much of the walk, even though I believed that most private owners had no objection as long as I respected their property and their personal privacy. Hence, if going the route involved getting arrested once, it was just another obstacle to be faced.

We continued on through most of the neighborhood, when we came to a channel which we had not anticipated. Normally, we would have swum it, and probably walked out the other end unimpeded. But Sue happened to be carrying a good camera, which we did not want to risk getting wet, so we started walking along the edge of the channel, toward the actual houses.

We saw that we were being observed from a pier, but walked toward the viewer intending to ask the best way out. But, he—a guard good at his job—waited until the closing distance had shrunk to the right spacing, then pounced like a fox seizing two chickens foolish enough to walk right up to his hole. Although he seemed torn between pride in the successful execution of his duty and confusion that we were a somewhat docile middle aged couple, he summoned his superior by radio. Together they called the local police. They evidently had a rule not to send out only one cruiser, so two arrived, making four in all.

"Shall I take a picture?" Sue whispered.

"Are you crazy?" I responded, our sense of "daring do" reversed.

"Just kidding."

Warned not to return, we were driven back to our car, without charges, in the back seat of a cruiser. A few weeks later, I called an acquaintance in the neighborhood, and he arranged for us to return to the area—through the gate—in our own car.

All told, this experience was not as frightening as a previous encounter with an unimpeded Doberman on the shores of Buzzards Bay. We stopped dead in our tracks on the beach as its bark boomed out from the bluff above, then proceeded to walk rapidly back from whence we had come. When it became clear that the dog was well trained to stay on the bluff, I suggested that Sue take its picture. This time it was she who would have no part.

Two thirds of the 427 miles of shoreline on Cape Cod are privately owned, necessitating a considerable amount of trespassing to walk around.

It is only due to a historical anomaly that the public in Massachusetts does not have universal access to all beaches. In most states, the zone between the tides is still publicly owned. However, when Massachusetts was a colony, private ownership was considered necessary to facilitate economic development of the shoreline. Hence, the Colonial Ordinances of 1641-47 granted the area between high and low tides to the owners, except for the purposes of "fifthing (fishing), fowling, and the paffage of boats (navigation)," for which the public retained rights of access. Ironically, although the ordinances were passed to promote economic activity, they are now being used by private owners to prevent such activities as aquaculture.

Ever since the Ordinances were enacted there has been legal sparring over their meaning. Aquaculture is not "fishing," but according to a recent opinion of Massachusetts Attorney General Scott Harshbarger, bird watching is "fowling." Harshbarger goes on to note, however, that "this issue has not yet been addressed by the courts."

The reality that laws cannot be changed retroactively should not be used as a reason not to change them at all.

Of course, anyone can wade one inch below the water line at low tide, but I am not sure I would try it in some neighborhoods if there is a security guard watching.

In addition to neighborhood associations with gates and security guards, there are a growing number of "no trespassing" signs. They grow thickest next to public beaches, undoubtedly to reduce the "trickle over" effect. Some signs, while pointing out the private ownership, welcome through walkers. Some use varying degrees of emphasis—such as "Police Take Notice"—to discourage them.

Fortunately, such negative encounters were far more than offset by the good will with which people escorted us through their neighborhood when we called ahead. There were also some genuinely pleasant chance encounters with people—on their property. I particularly remember a very friendly conversation on a property which jutted into the harbor atop a wooden bulkhead in West Falmouth Harbor, after my daughter and I came face-to-face with the owner on the edge of his lawn. He was weeding a garden on his hands and knees. It was one of the few times when we clearly had violated an owner's privacy.

In most tourist economies the shores are more user-friendly. I found an indication of this last summer—while vacationing in Bar Harbor, Maine. It was a sign which read: "Shore Path, C1880, For more than 100 years this shore footpath has been open to public use by the generosity and cooperative spirit of the owners of the several properties through which it passes. Please respect their privacy and enjoy the beauty of the historic waterfront. Misbehavior by individuals or parties, and misuse which results in damage to the path, to features along it, or to private grounds it borders are sufficient cause for closure to public use."

Indeed, in most coastal states the entire beach between high and low tides is still publicly owned. A friend who recently visited Virginia Beach, told me that when he asked where he could go on the beach, the respondent named the northernmost point on the Virginia coast and the southernmost point, and answered "anywhere in between."

In contrast, residents in many bedroom communities closer to Boston have done far more to deter public access than have residents on Cape Cod. In some, the simmering anger on both sides has produced much legal expense but no winners. Unfortunately, this seems to be the direction in which the Cape is headed.

Although I encountered a few signs on the Cape similar to that in Bar Harbor, our shoreline is not user-friendly in some of the places where it most needs to be, such as downtown Hyannis and Orleans. Even Boston now extracts participation in a harbor front sidewalk as a trade-off for new development in the area. If we are going to continue to expect "guests" (paying and otherwise) to come to our shores, we need to improve access where they most want to be—by the sea.

Monomoy Island, Chatham
Other Summer Visitors

It was a warm June night, the tide the highest of the month. The water's edge was lined with horseshoe crabs. The crabs were engaged in an annual orgy, laying eggs at the high water mark to hatch a month later when the next similar tide recurred. The males, smaller than the females, had hitched a ride attached to the back of the females so that they would be dragged over the eggs to fertilize them. The night culminated an annual migratory journey of 50 miles or more from the mucky bottom at the edge of the continental shelf, where the crabs had hibernated through the winter.

The horseshoe crabs had returned to breed, feed, and summer on the Cape. Like most of nature's creatures, they came to eat other species, knowing just when and where to find a very tasty meal.

The human population of Cape Cod triples on a busy summer weekend. But, we are not the only ones to utilize the Cape's attractions for a summer visit. Scores of birds, fish, reptiles, and mammals also stop by. Many are on view to walkers of Cape Cod shores.

Horseshoe crabs were undoubtedly among the first visitors to Cape Cod. They have survived for 350 million years, preceding dinosaurs for over half of that period. In fact they are not crabs at all, but anthropoids, members of the same family as spiders and scorpions. Horseshoe crabs are extremely efficient mechanisms; they have to be to have survived for so long. Their bodies contain only the few parts basic to their survival, nothing more.

The crab's hard shells are shaped like Buckminster Fuller's geodesic dome, reputed at one time to be the most efficient combination of minimal weight and maximum strength ever developed. Their spiny tails serve as a lever to right themselves and to push off. They have ten eyes for varying purposes and in varying locations—five under the shell. Would it not be nice if we could see behind and below at the same time as ahead? Their bodies are comprised primarily of five pairs of legs, which also assist in the feeding process, although they can survive for a year without eating. The stomach is conveniently located amidst the base of the legs. A row of book gills—fleshy flaps—serve as paddles for swimming.

Horseshoe crabs possess two attributes that have been of invaluable use to mankind. Their shells contain Chitin, a substance which speeds the healing process. Man has now synthesized it for utilization in artificial skin. The blood of a horseshoe crab has a unique ability to detect the presence of bacteria in human blood and is utilized in tests for that purpose. Cape Cod horseshoe crabs are caught, trucked to a lab in Woods Hole, bled, and returned to the sea. The blood sells for thousands of dollars a quart. Horseshoe crabs thus represent a prime example of the very real fear that every time a species disappears from earth, important substances of potential use to mankind may go with it, a belief that is one of the cornerstones of the Endangered Species Act.

While horseshoe crabs have not been classified as endangered under the act, they may be seriously threatened by man. Shellfishermen have always had a reason not to like them—they eat popular shellfish. For a long time kids were paid a nickel bounty for every tail they brought in. Another active industry chops up the "crabs" for fishing bait.

Although the crabs are returned to the water after being bled in Woods Hole, their survival rate is unknown and under debate.

Monomoy Island, Chatham

The remaining horseshoe crab population in the United States is estimated to be about one tenth of what it once was. Calls grow for more research and regulation of horseshoe crabs. But until more is known, the future of the oldest of Cape Cod visitors will remain in doubt.

The Monomoy Islands south of Chatham are the place on Cape Cod perhaps most closely identified with many species, endangered and otherwise. The islands represent the first land many birds and sea dwellers reach as they return from the south, and the last staging point they visit before heading back for the winter. Monomoy is thus Cape Cod's Ellis Island of sorts. It is an ocean landing strip and a welcome sea port all wrapped into one. Like nearby South Beach in Chatham, the islands are a product of the ever shifting sand washed down from as far north as Eastham by the constant erosion of the Cape's shoreline. Prior to a breakthrough in 1978, the islands were one single land mass, and, until a similar break in 1958, they were attached to the mainland at Morris Island in Chatham. Presently accessible only by boat, they are one of the few places on Cape Cod where another species—seagulls—are perceived to be a bigger threat to life than man.

The islands were taken by the federal government in 1944 and designated as the Monomoy National Wildlife Refuge under the Migratory Bird Conservation Act. (During World War II, even after the 1944 designation, Monomoy was used as a gunnery range by the military.) In 1970, the islands were also designated as the Monomoy Wilderness to be managed in accordance with the Wilderness Act of 1964. All told, those responsible for the islands thus must consider three statutory objectives, conservation of migratory birds, preservation of the "wilderness character," and, under the Endangered Species Act of 1973, "conservation of endangered and threatened species." Three hundred and five of the 453 species of birds ever seen anywhere in Massachusetts have been recorded at Monomoy.

There has not always been universal agreement concerning

how the Refuge should be managed to fulfill its triple purpose. Of primary concern over the past two decades has been what to do about the burgeoning seagull population. Seagulls and other birds and animals have been killed at Monomoy in order to make the environment safer for rarer species, but inevitably the results have never been as fully intended. There are too many complex interrelationships between too many species to facilitate any quick and easy solutions.

Walking Monomoy provides a variety of unique experiences. The relative seclusion and abundance of wildlife make the islands a marvelous place to walk, view nature, and gather one's thoughts. Except for private boaters in the summer, the islands can only be reached by a limited number of naturalist guided tours provided by the Cape Cod Museum of Natural History and the Audubon Society. Rough weather cancels approximately half of the tours planned. On my most recent visit, there were just six of us on South Monomoy, the normal limit for a Museum tour.

The sense of peace and solitude are astounding. Seagulls are in charge, and they know it. Their squawks provide continuous background music, drowned out only by the roar of the surf on the ocean side. The seagulls seem fearless; it is possible to get much closer to them than on the mainland.

Multiple signs restrict access to the prime nesting areas on much of the island. Although this is probably necessary and desirable to maintain an undisturbed wilderness area, the signs and the limitations which they represent elicit a variety of emotions and thoughts. Prison camp? Zoo? One comes to Monomoy to feel free and the mind does play games as the eyes roam the vast expanses. Are the signs intended to keep people out or wildlife in? Nothing is ever totally natural. For example, two freshwater ponds, complete with nesting islands, were dug out to attract migratory waterfowl.

While viewing deer, owls, and other birds through binoculars in one large expanse of restricted land, I was reminded of the movie, Out of Africa. I truly would not have been surprised to see an elephant or giraffe walking around. Monomoy has that effect on you.

I walked the beach for about five miles around the south end of the island. It was a walk through several different worlds. As I started out two miles up the coast on the east (ocean) side, the beach was very rough, wide, and coarse—culminated by a high and ever-changing line of dunes. On the west side, the beach is smaller and gentler, and the sand as fine as any I have seen.

Even more striking were the changes in the weather. It was sunny as I started. But, by the time I reached the end of the island, it was totally foggy. As I walked back along the west side, the sun gradually reappeared. I could have measured my progress by the amount of fog even if there had been no landmarks. This was not unusual, the area is known as the "fog factory."

One of the most distinguishing features of the Monomoy and Chatham area is fog. Advection fog occurs when warm moist air comes in contact with a colder land or water surface. When chilled, the moisture in the air condenses into droplets of fog. Old timers look to humidity as a precursor of fog.

A considerable amount of warm moist air comes into regular contact with a colder surface in the Monomoy vicinity. Well offshore, the warm water Gulf Stream and the cold water Labrador Current meet, block each other, and travel off side-by-side into the north Atlantic. Closer to home, Nantucket Sound on the west side of Monomoy contains warmer water than the ocean water to the east. Finally, warm moist summer air ends up its passage across the entire country by exiting over Cape Cod, where it meets the colder ocean surface. During the summer there is almost always some type of warmer air coming in contact with a colder surface just off Monomoy. *Weather for the Mariner*, written by Captain William J. Kotsch in 1970 for the Naval Institute, shows a fog bank characterized by 20 to 30 days of fog in the June to August period starting at Chatham and extending to sea just north of the Gulf Stream between the 60 and 65 degree temperature lines. The fog bank extends slightly north or eastward, passing to the south of Newfoundland before bending sharply northward to the east of that island.

Once fog has formed over the adjacent ocean, all it takes is a south wind to blow it in over Chatham and Pleasant Bay. That, too, is almost automatic. On a hot summer day, the cooler air from the ocean is drawn in underneath the warmer air over land, fog and all. Chatham fog even has a name, a "smoky sou'wester."

Everyone has a favorite fog story. In his book, The Bay–As I See It, W. Sears Nickerson says that he has seen "not once but a hundred times" fishermen leave Chatham on a foggy day, navigate and fish far off shore, and return from the fog to "hit the harbor right on the nose every time." My friend, Bruce Hammatt, tells of returning one time from Nantucket in the fog, buoy to buoy, by compass. "Compasses really do work," he said. As he reached Monomoy, a power boat appeared in the fog, asking the way to Hyannis. Hammatt pointed out a visible buoy and instructed the skipper what compass heading to take from the buoy. The boat went to the buoy, then, obviously not knowing how to read a compass, set off in the exact opposite direction.

My favorite fog story occurred inland—over and over again. When my daughters were young we would travel from Brewster to Chatham several evenings each summer for recreational softball games. As often as not we would leave a sunny evening halfway in between, play the game in the fog in Chatham, then find the sun about where we left it as we returned to Brewster.

As I walked along the Monomoy shore, I came upon a group of approximately 40 gray seals, piled upon one another on the shore like a stack of bean bags. Although I did my best not to disturb them, one by one they slipped into the water. When I walked to the water's edge, there they were, slightly offshore, peering as curiously at me as I at them, seemingly knowing that while I may have commanded the land, they certainly commanded the sea. As I continued my walk, they followed along for half of a mile or so.

The walker of Cape Cod cannot help but observe the interrelationships between the many species. Interesting stories could be told of many of these species, including the cod fish—after which the Cape is named—or of the oyster or scallop, popular symbols of the Cape. Four stood out to me, the horseshoe crab, the striped bass, the osprey (which I will consider in more detail later), and the piping plover. All are highly visible to a walker all of the way around Cape Cod. None are uniquely connected to the Monomoy Islands, but each have had a presence there.

Whereas the horseshoe crab population is clearly in decline, striped bass and osprey have rebounded. The greatest threat to the continued existence of both was the same—the pesticide DDT. Striped bass winter and spawn in Chesapeake Bay, which became severely polluted. The Bay was cleaned up and now the population of striped bass is as large as ever. Ospreys absorbed the DDT in fish they consumed as the chemical moved up the food chain. Both striped bass and ospreys were once considered endangered.

Striped bass are a uniquely coastal fish—they feed by chasing schools of smaller fish into shallow water where they can be consumed in the ensuing confusion. No striper has ever been seen more than three miles off the coast. All the Cape's bass pass by Monomoy between May and November on their way between Chesapeake Bay and the Cape's other waters.

The feeding habits of stripers make them uniquely popular with sport fishermen, who are visible surfcasting all of the way around the Cape. Inasmuch as bass are very tasty, both commercial and recreational boats pursue them as well. Indeed, striped bass fishing is popular enough to be a significant boost to the local economy. Nationally, the value of the striped bass fishery approaches $1 billion; the near total loss of the fishery for many years while the population was restored was thus a serious economic as well as recreational blow to places like Cape Cod.

Recent management of the striped bass fishery represents a substantial success story. Stocks are now thought to be as great as they were before the crisis was recognized, having rebounded to over ten times the number at which they bottomed out. Since the number of fish being taken remains only at approximately twenty percent of what it once was, the minimum size for a keeper was recently lowered from 34 to 28 inches. Initially the reduction was proposed for both recreational and commercial fishermen, but after protest, the minimum was restored to 34 inches for commercial fishermen. No one wants to put the striped bass back at risk again.

The Endangered Species Act of 1973 was another of those blockbuster pieces of legislation which has probably been used for purposes which were never foreseen by those who enacted it. It states that "all Federal departments and agencies shall seek to conserve endangered species and threatened species and shall utilize their authorities in furtherance of the purposes of this Act." Despite hundreds of pages there are no ifs, ands, or buts, no ways/means test, and only one exception—for national security. In a recent example of the Act's reach, the aversion of just one plover family on Kalmas Beach to the sound of fireworks was cause for a substantial hullabaloo over whether the traditional Fourth of July fireworks display could be held in Hyannis. The controversy ended when the Audubon Society, perhaps realizing that more harm than good would be done to their cause if there were no fireworks, arranged for a barge to be used to launch the fireworks. The presence of several species of moths threatened the proposed site for a new prison on the Massachusetts Military Reservation. No humans want the prison in their "back yard," neither do the moths.

It would be interesting to see what the call would be if one endangered species were found to be eating another.

The current poster child—and bete noire—of the endangered species movement on Cape Cod is the threatened (a lesser category under the act) but not endangered piping plover. Small birds, seemingly omnipresent on Cape Cod beaches despite their threatened status, piping plovers have the unfortunate habit of laying their eggs right in the middle of the beach at a time of year when humans also want to be there.

Monomoy Island, Chatham

Public authorities have felt compelled by the act to keep a close eye on beaches for piping plover nests, designating large segments as off-limits or fencing them in to protect the hatchlings. The closings have resulted in considerable conflict, particularly with the operators of four-wheel-drive vehicles denied access, including striped bass surfcasters who have found it more difficult to get to their prey. Frustrated humans have been known to deliberately destroy plover eggs in a dispute which may never be totally resolved. The plover population has rebounded between 1989 and 1996 from 4 to 20 on Monomoy, 16 to 97 in the Nauset-Monomoy system, and with similar increases elsewhere on Cape Cod. But, as long as they lay their eggs in harm's way, they may require some form of protection.

Piping plovers look like many other shore birds, sanderlings in particular, which are more adept at staying out of harm's way. To the best of my knowledge, since use of their feathers in women's hats was prohibited in 1918, they have possessed no commercial value to mankind. But, by my standard of values, fishing is a sport which should be engaged in by boating or walking. For that matter, four-wheel-drive vehicles—although I own one—do not enhance the pleasure of beach walking; perhaps walkers are the endangered species.

Extinction is forever. But, comparing four endangered species (piping plovers, ospreys, striped bass, and right whales) and one species that is not considered endangered (the horseshoe crab), humans certainly could fine-tune our ability to compare the relative value of the consumption of resources and human sacrifice against the perceived gains from protecting endangered species. Indeed, at this moment in time the horseshoe crab may be more endangered than striped bass, ospreys, or piping plovers. Wouldn't it be something if we kill off the Cape's most long standing visitor, one of nature's most perfect creatures?

Ospreys eat striped bass carcasses left by fishermen. Piping plovers obstruct fishermen from four-wheeling to their prey.

Birds engorge on the eggs of horseshoe crabs and striped bass. Furthermore, striped bass are not growing as fast in size as they once did, perhaps because—in the all-or-nothing world of endangered species—no one has paid as much attention to the numbers of the fish stocks which the bass consume as they do to the stripers themselves.

The life cycles of living species are interrelated. Mankind is improving its ability to share space. But, we need to continue working at it.

Chatham Harbor
Man Against the Sea

For much of the past ten years, the scene has resembled a war zone. Large chunks of pavement have been strewn at random. Bricks and concrete blocks, once part of house foundations, have been piled up on the shoreline. Twisted pieces of pipe have protruded from the ground. Most startling of all have been the remains of trees. Until cut down, some have stood like frozen soldiers in the sand, their exposed roots taking the form of eerie sculptures.

While the distance between Andrew Harding's Lane and Holway Street on Chatham's shifting shoreline is only about an eighth of a mile, the never-ending struggle between man and ocean is perhaps more apparent there than any place on Cape Cod. After a remarkable 1987 winter storm cut through Chatham's outer barrier beach, revetments were built to protect houses to the north and to the south. But, until 1996, no revetments were built between Andrew Harding's Lane and Holway Street, leaving this vulnerable stretch of sand—a "no-man's land" of sorts—exposed to the full force of an open sea with an appetite of house-sized proportions. To date, ten homes have been consumed or condemned. More face an uncertain future.

Today, the pounding continues, as the ocean seems to be earnestly at work at creating a new cove, and has pushed the shoreline back more than 100 feet for much of the stretch. Prior to construction of the 1996 revetment, during a stormy high tide waves continued to lap at the foundation of an eleventh house. They even spilled into backyards further up Holway Street until construction of barriers in 1997. Sand that has washed into the neighborhood stands high and dry like an inland sandbar.

The Bible tells us that a thousand years is but a day to the Lord. It promises to be an interesting week and a half if geologists are correct that the entire Cape will be eroded away in 10,000 years. Sometime within that period every Cape neighborhood—street by street—will undergo the tragic havoc wrought on Chatham for the past 10 years. It would be interesting to see the difference in public attitudes if it were happening everywhere at once. Still, every cloud has a silver lining. If we all hold onto our property, our descendants will all eventually come to own waterfront property. And if everyone has good timing, it might even all be sold to wealthy foreigners at just the last minute.

Comparatively speaking, the Cape is a uniquely masochistic place, in that much of its population seems resigned to its ultimate demise. Many Cape dwellers almost seem to welcome such a cataclysmic event—after their own death—perhaps to purge themselves of some such sin as having been too materialistic or for wresting the place away from the Native Americans to begin with. New Jersey, Florida and other shore-front states are fighting much harder for their continued existence. The U. S. Army Corps of Engineers is in the middle of a $1.7 billion, 50-year program to restore 21 miles of New Jersey shoreline with sand pumped from the ocean floor. When an effort was made to end such federal largesse, a Democratic New Jersey Senator and a Republican from Florida led its defense. Pork barrel in America crosses partisan and ideological lines.

A walker can see it all on the Chatham coastline. Coming around the corner from Nantucket Sound to Chatham Harbor, Morris Island has been protected primarily by large bags of sand. Next comes a massive example of the ocean's capability to move sand. A new tombolo—a massive sandbar made up of the sand from the breach—now connects the mainland at Lighthouse Beach to the diminishing southern portion of the barrier beach, thereby for the time being dividing what was one estuary into two bodies of water. Then come numerous revetments surrounding both ends of the gap between Andrew Harding's Lane and Holway Street. Some were built before 1987, most afterward.

In 1996, a new revetment was built at town expense to protect Holway Street. It fills only half the gap, and by bending and becoming perpendicular to both the adjacent revetment and the shore in the remaining unprotected portion, seemingly puts houses further up Holway Street and at the Andrew Harding's end at even greater risk, leading me to wonder whether the "experts" knew something that I did not.

The scene at the end of the new revetment remained eerie during high winter tides of 1996-97, as waves lapped under and through what had been live trees on the shore until the unusually high tides arrived. Construction of the revetment also underscored the continuing inconsistency and inequity of the public response to the havoc caused by erosion. The new revetment was of considerable potential financial benefit to one owner whose house was saved, another owner of the partial remains of a condemned house which could now be restored, and a third owner of a partially restored lot. In contrast, the owners in the remaining unprotected areas became, if anything, even more threatened.

Residents initially sought as a group to install "rock mattresses," a layer of lobster trap like cages filled with small stones, along the shoreline. Skeptics questioned the likely effectiveness, but the plan was abandoned anyway when the cost came in at over twice the estimated $25,000. The price had been driven up by the winter storms. In the summer of 1997, a "heavy duty" plastic fence which looks like a bulkhead was built behind two more houses on Holway Street, also perpendicular to the original shoreline. It continued to leave much of the new cove exposed as the unprotected section of the shoreline moves ever-closer to

Main Street in Chatham. Viewing the Chatham shoreline thus confirmed the conclusions I had reached elsewhere on the Cape. Hard solutions work, but just for a while. No houses protected by revetments have yet been lost in Chatham. Soft solutions are ineffective, at least against the full power of the open sea. The sand bags at Morris Island, as well as elsewhere, lie in disarray. I am tempted to call such solutions frauds, but will settle instead for the term "placebos." They make it easier for both those who oppose hard solutions and property owners to think that they are doing something to protect property.

Revetments, by maintaining shorelines further out than they would otherwise be, may result in the loss of beaches. I passed revetments in Chatham which left me, even at low tide, with the unhappy choice of scaling the revetment rocks or sloshing in the knee-deep water below them.

It thus comes down to the choice of protecting properties for a while—perhaps for a human generation or so—versus loosing some beaches. For a benevolent despot (or any other single rational individual) the decision might be easy. Build revetments—without gaps—in the most densely populated areas through a betterment type process. To protect the beaches, allow nothing elsewhere. But, Chatham is not governed by a benevolent despot and neither is the rest of the Cape. Instead, we have four changing and frequently conflicting government levels in our democracy.

To date, neither side has been a winner in the struggle over whether and how to combat erosion in Chatham. Ten houses have been lost and others threatened—the lives of their occupants irreparably altered. The parking lot, beaches, and easy access at what were once two convenient landings in the middle of town are gone as well.

Many might thus consider Chatham a symbol of the failure to plan, the failure to get along, the failure to have expeditious legal and political procedures to work things out, and of the inconsistency of the "winner-take-all" nature of the procedures which do exist. The band plays on.

Chatham Harbor

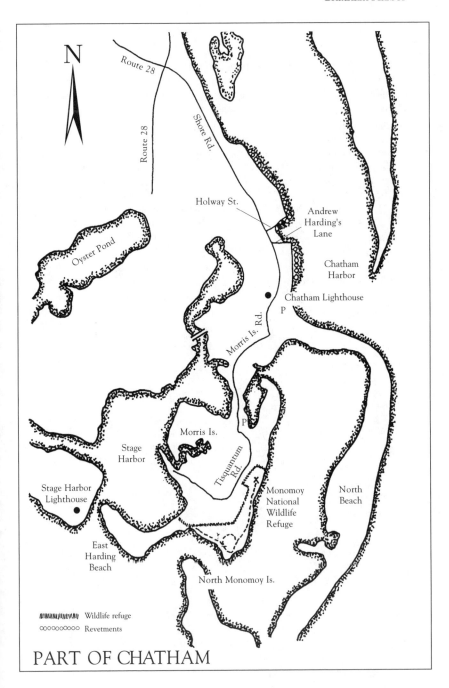

PART OF CHATHAM

WALKING CHATHAM

The walk around Morris Island from Morris Island Road to the Monomoy National Wildlife Refuge offers a variety of attractions. Initially, the walker passes between Chatham Harbor (with a view across to the South Beach) and the upscale housing of Morris Island. The route passes the remnants of efforts to fight erosion with large sand bags. After turning the southeast corner of Morris Island, there is a designated loop trail in the Refuge, with North Monomoy Island visible across the water to the south. At low tide it is possible to continue beyond the refuge along the ever-changing shore of East Harding's Beach to the Stage Harbor Channel, with a view of Stage Harbor Lighthouse on the far side.

The new (since 1987) tombolo (sand spit) in front of Chatham Light leads to South Beach, another interesting walk previously inaccessible by foot from the mainland. Although this walk may currently (1997) be undertaken at any tide, conditions are variable.

The area slightly to the north of Chatham Light, between Andrew Harding's Lane and Holway Street, provides a dynamic view of erosion and the impact of revetments. It is best reached down Andrew Harding's Lane at low tide.

Pleasant Bay
Powermen & Drifters

My friend Andy Young says that "you can see more, do more, and learn more in the short trip across Pleasant Bay than on five hundred miles of open ocean." The Bay is a constantly changing body of water, the weather is unusual, and human activity in the form of fishing, shellfishing, birding and boating is varied and interesting.

Among the birds on view in Pleasant Bay are the ospreys nesting near Strong Island. Ospreys, otherwise known as Fish Hawks or Sea Eagles, are another wildlife preservation success story, with the bird once again visible in may of the Cape's bays and marshes. They are large birds. Their bodies can be up to two feet long, their wing spans up to seven feet, their weight four pounds. They mate for life and live up to 30 years. They fly 25 to 30 miles-per-hour.

Ospreys soar approximately 50 feet above the water looking for prey—fish. When they see one they drop down in a controlled (steered) free fall (the fastest and quietest way of getting there) to strike their prey with a loud and decisive splash. Their talons each have two front and two back claws, the shape of one third of a circle, hinged, and extremely sharp. Kind of like Captain Hook with four hooks on each hand. Ospreys utilize their claws to pierce and tightly grasp their prey. After a successful catch, an osprey will hover just above the surface of the water for a moment to get control, then, if all goes well, rise like a helicopter with the fish attached and return to its nest. The birds have been known to drown if the fish turns out to be too big and they are unable to disengage their claws in time. Ospreys can fly carrying prey up to their own weight, which they hold like a torpedo below them to reduce wind resistance.

Prior to the widespread use of DDT, there were 90 known pairs of ospreys in Southeast Massachusetts. The number fell as low as 12, before rebounding to approximately 250. The rebound has been facilitated by what might be the greatest public housing program ever created. Ospreys dwell in large nests built considerably above the ground along the shore, traditionally in dead trees and other similar locations. But after shoreline development substantially reduced the number of suitable sites, Commonwealth Electric came to the rescue, placing telephone type poles in carefully selected locations in salt marshes, with pallets on top to serve as the base for the nest. Thanks to man's help, ospreys may well be better housed today than they ever were.

No one knows how Pleasant Bay got its name. But, everyone knows that it was not a bay at all until eroding sand began to wash down from the shore to the north after the last ice age, creating a new embayment between the ever-changing sand spit which forms its eastern shore, and the original shoreline to the west. The eastern shore of the Bay is known as North Beach and is part of the Cape Cod National Seashore in the towns of Chatham and Orleans. The Bay's western shore, or inland shore, is primarily privately owned.

Pleasant Bay is home to three rivers, two creeks, three coves, eight ponds, and one harbor—many of which do not fit the dictionary definition of their designations. I walked up the west side of Pleasant Bay to Namequoit Point in Orleans, then swam across to the end of Barley Neck. The Bay also has four points and two necks, along with several islands that are only completely surrounded by water during the highest of tides. Cape

Codders have always named things whatever they have wanted.

Pleasant Bay is perhaps most beloved for its boating. In 1996 there were 1,615 boats moored in the Bay, an increase from approximately half that number 20 years ago. Many more are brought to the Bay by trailer on nice summer days, when it gets crowded out on the water.

There are two types of boaters, powermen and drifters, otherwise known as sailors. (Their vessels are known as stinkpots and blowboats respectively.) Sometimes I wonder if ever the twain shall meet.

To own a boat, one must like to take care of it as much as to use it. I have always found it easier to go out in friends' boats. I have many friends who are sailors, and many friends who like powerboats, but none who will admit to both. There is no argument between them, just a stark contrast. As Bruce Hammatt puts it, "a boat is vehicle to an activity." Powerboaters use their boat to do something else—fish, picnic, water ski, sightsee—for which or to which the power boat is a mode of transportation. When I asked one friend why he prefers a powerboat, he responded, "I have to be ready to go to work on Monday."

Sailing, on the other hand, is the activity. Sailors commune more with nature. One calls himself "the harness between the wind and the water," another says he seeks the journey, not the destination ("I do not want to get there"), and a third says he sails for the peace and quiet, "to get away from the stressful and noisy mechanical world which we live in." One friend rarely even uses his sailboat. He just uses it as a place to go out on the water and sit.

Sailors also consider themselves more cerebral. To get where one wants to go one must know the winds and the tides and overcome them. This is particularly difficult in parts of Pleasant Bay where the channel is narrow and the current strong. If sailors are not skilled in such instances, currents can quickly send them back from whence they came, or even aground.

When I was a boy spending one month each summer on the

western shore of Buzzards Bay, I happened to be in between two well-populated age groups of childen, one including my older brother, the other my younger brother. But there were few kids my age. As a result I had a tendency to get squeezed out of events such as sailing races—the times when most youngsters really learn to sail. My older brother would enter the "junior" category, and my younger brother the newly created "midget" category. I was relegated to crewing for anyone who would have me, a category which did not include either of my brothers.

One day a friend, whose older sister usually crewed for him, asked me to join him. I do not believe that the chemistry in their boat had been very good either. We got off to a poor start at which time, always the contrarian, I advised him that since every other boat had gone off on a starboard tack, we might as well go the other way. If it worked out, we might win, and if it did not, everyone was going to have to wait for us to finish anyway. I viewed it as a fifty-fifty proposition versus one representing no chance at all.

We won, which gave me a reputation which I knew I could not live up to. The only ability which I had demonstrated was the youthful realization that in the tides and currents of human affairs, the pack is as likely to be going in the wrong direction as the right one. I retired from sailing for a year or so, but my friend grew up to buy the local boat yard and run it for many years.

To balance the record, I also have my own powerboat, or perhaps I should say "fishing" story to tell. One evening many years ago, obsessed with the idea of catching a fish, I went out in the boat by myself. When I finally hooked one, which turned out to be below minimum, I was astonished to realize that I had hooked it through the eyeball. (I am told some varieties of fish will eye the bait closely, at which time a slight movement may hook it.) Despite valiant efforts, when I extracted the hook the eyeball came with it. When I threw the fish back, it circled the boat several times before departing, leaving me to wonder whether it could still swim straight, was looking for its eye, or just

Pleasant Bay

did not want to let me and my guilty conscience off the hook.

Not all powerboats or sailboats are the same. Many very skilled seamen make their living on a powerboat. At the other extreme, one captain of such a working powerboat told me that he "had never seen a jet skier who was acting responsibly."

Often there is a daily cycle to the flow of boats on Pleasant Bay. In the early morning work boats dominate, trying to get their chores done before the wind comes up. As the day progresses, recreational boats and sail boats take over. Toward evening, particularly if the wind dies, the work boats are out again, along with a few sailors trying to relax after a day at work.

As I passed through Pleasant Bay on my walk, I decided to go out for a day of sailing with one friend, and a day of power boating with another. The pleasures of both were in abundance. Although we did not go anywhere in particular or do anything of significance, the sailing was peaceful and relaxing. As we returned in the early evening, wind dying, I shall never forget the soothing sound of the boat lapping gently against the water of the quiet evening.

In contrast, my audible memory of a recent powerboat ride is of the wham, wham, wham of its hull slamming against the wake of other passing boats. Still, in the course of just one morning, we took photographs of Chatham's eroding shore, cruised down to Monomoy Island to view seals, and tended to the owner's six lobster pots.

Both sailboats and powerboats have their time and their place. But, it would seem, never the twain shall meet.

WALKING NAUSET BEACH AND POCHET ISLAND IN ORLEANS

There are three interesting walks which may be made from the public parking lot at Nauset Beach in Orleans.

A walk along the beach to the north leads to the inlet through Nauset Beach into Nauset Marsh as well as the Town Cove and Nauset Harbor areas of Orleans. To vary the route, the walker can go along the beach in one direction, and along the bluff below Nauset Heights in the other, although this route can

Walking the Shores of Cape Cod

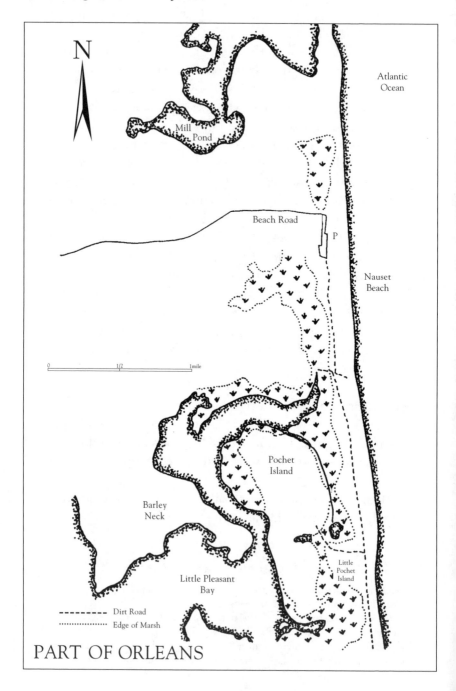

be obstructed in the spring and early summer by nesting shore birds.

By following the beach road south out of the parking lot approximately a mile and a half, a walker comes to a road and bridge to the west leading to Pochet Island. There are private residences on the island, but walkers are welcome if they obey several restrictions listed on a sign. There are no paths along the shore around the island, but the shoreline—with views of Little Pleasant Bay—is accessible for the adventuresome. There are several paths through the dense woods which covers much of the island. To the south, the experience walking on Little Pochet Island is much the same. Both islands are only true islands when high tides flood the salt marshes that separate them from the remainder of the beach.

North Beach extends for several miles south from the Nauset Beach parking lot. Many viewers come to the beach following an ocean storm to see the wave action. There are a number of beach cottages near the end of North Beach at the celebrated "North Beach Cut" caused by a 1987 storm that now separates North and South Beaches.

Town Cove, Orleans
Planning

One property on the downtown side of Town Cove in Orleans has asphalt paving dumped down the bank behind—just another method of trying to stop erosion. Open discharge pipes ooze liquid from under several nearby buildings. There is no odor to imply that any is septic in nature, but the slope behind one is discolored into a rusty orange shade. Debris of various descriptions litters the backyard of several other properties in winter when there is no marsh grass to obscure the view. It is unclear whether it floated in and was never removed, or merely left to rot. Some of both, I suspect.

The two sides of Town Cove in Orleans are more like two different worlds than merely two different sides of a cove.

The eastern shore, like most of the Orleans waterfront, is elegant housing—some exhibiting the charming character of age, some the brash arrogance of youth. An informal dirt path along the water's edge testifies to the neighborhood's popularity and togetherness, as well as to the waterfront's use.

The opposite shore cannot be described in equally complimentary terms. Although the body of water is the same, the character is not. This neighborhood is neither "fish nor fowl" but rather a mixed salad containing small seasonal cottages, a few larger houses, two sizable boat dealers, a large number of homes transformed into retail outlets, and assorted other properties.

Why is such a potentially attractive stretch of waterfront left shabby in places? Looking back, it is probably the result of zoning mistakes. Looking ahead, it is a classic example of the current inability of government to push a car stuck in the mud out, either backward or forward. If it were Boston, the old Boston Redevelopment Authority would have taken the entire area by eminent domain and sold it to some developer to raze and rebuild, thereby improving the tax base. But, on Cape Cod there is no such thing as the Cape Cod or Orleans Redevelopment Authority. Instead, there are ever-changing Boards of Selectmen and a host of other town committees, mostly voluntary and some with conflicting purposes and goals— a balance of power if one views the glass as half full, a self-fulfilling formula for inaction if one sees the emptiness. Government today has a committee for every constituency and as a result, very little capability for collective decisiveness.

Most of this neighborhood was once residential. But, with the advent of zoning, much was designated either limited or general business. As one Orleans old-timer put it, "at the time everyone believed that commercial zoning would ultimately provide for the enrichment of us all." Astonishingly, now much of the area is focused on Route 28 in front, not on the water behind—yet another example of the never-ending lust to stop cars in order to sell whatever might be bought. The water behind is rarely even visible to passers-by.

Unfortunately, too much of previously residential downtown Orleans was similarly zoned, creating a sprawl which exceeds demand. To cap it off, two shopping centers were built on both ends of town, thereby eliminating the need or desirability for many shoppers to even come in to the central business district.

As a result, no one has invested much in either improving the residential properties or in building attractive business facilities in what could be, for example, a marvelous location for a classic waterfront restaurant or hotel.

What should be the most attractive part of downtown

Orleans is not. Imagine how much more attractive Orleans—or Hyannis—would be under any one of three models: a park along the water with an open vista like either side of the Charles River in Boston, an attractive redeveloped waterfront business area like that along Atlantic Avenue in Boston, or an attractive residential neighborhood—hopefully with a rustic, but public, nature trail—like that on the other side of Town Cove in Orleans.

But, many in Orleans do not support a larger or more attractive downtown business area, and many in the business community oppose any retreat in their zoning. As a result, downtown Orleans may remain "stuck in the mud." Hopefully, that will not always be the case.

Orleans is just one example. Unfortunately, these problems are not unique. There are other places on Cape Cod where planning and zoning are similarly stuck, unable to move in either direction. There needs to be a way around such impasses.

Cape Cod needs a vision and a plan for its shoreline.

The plan must recognize that the attractiveness of our shoreline is just about the only reason why anyone comes to Cape Cod as either a visitor or a resident. It is also therefore related, either directly or indirectly, to all efforts to make a living on Cape Cod. The natural beauty of the shoreline must be preserved, must remain as accessible as possible to all, and must be utilized in places for economic purposes.

Until there is broader agreement concerning long-range planning for Cape Cod's shoreline, there will continue to be disagreement concerning every proposed change.

WALKING NAUSET MARSH

Nauset Marsh, made famous by Henry's Beston's The Outermost House, and Wyman Richardson's The House on Nauset Marsh, is rich in both history and natural life. It is easily accessed for walking at three spots:

• The Cape Cod National Seashore provides parking and excellent trails at Fort Hill in Eastham and overlooking the marsh and through a nearby maple swamp. When walking the Fort Hill Trail around the perimeter of

Walking the Shores of Cape Cod

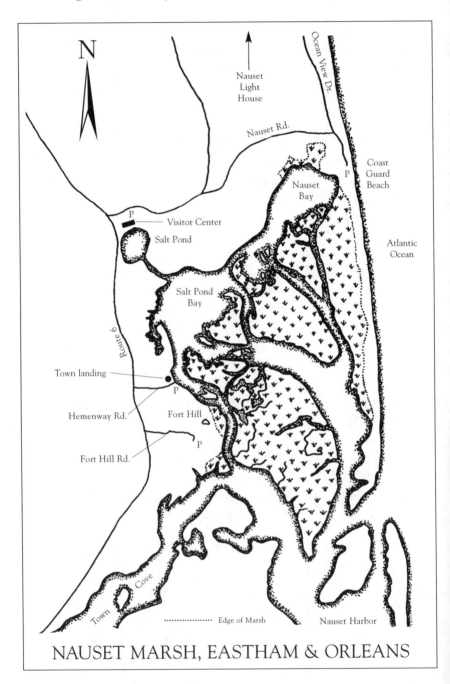

NAUSET MARSH, EASTHAM & ORLEANS

Town Cove, Orleans

the hill overlooking the marsh, follow the signs to Hemenway Landing in Eastham. Walk the path along the shore for as long as possible as an alternative to the trail above.

- Trails from the Salt Pond Visitor Center in Eastham lead to Coast Guard Beach and along the shore around Salt Pond.

- From Coast Guard Beach it is possible to walk north along the beach to Nauset Light and beyond as far as one desires towards Race Point in Provincetown. Walking south, it is possible to walk along the beach to the channel which separates Nauset Marsh from the open ocean. Like so many barrier beaches, this walk provides excellent views of the ocean on one side and the marsh on the other.

Nauset Light, Eastham
Cussing the Moon

Watching a lighthouse roll from place to place should be a once in a lifetime experience. Nevertheless, in 1996, two of the fourteen lighthouses remaining on the Cape Cod shore—Highland Light in Truro and Nauset Light in Eastham—were moved to prevent them from falling into the sea, as nearby cliffs eroded away.

Nauset Light was moved in November. Jacked up onto the back of a wide, over-sized flatbed truck, it was moved ever-so-slowly out to the road, along the road for a short distance (while utility crews moved wires), then off into the woods to its new home where it should be safe from erosion for the next century or so. Crowds pressed behind the ropes on each side to get a glimpse, as if the lighthouse were some famous celebrity. What, I wondered, would happen if the driver mistakenly stepped down too hard on the accelerator?

What is it about lighthouses that makes them so unique that people want to move them back while they are willing to watch other structures fall into the sea? If geologists are right about the entire Cape disappearing in 10,000 years, will all 14 lighthouses be clustered together in the final days on the last remaining bit of dry land?

Nauset Light played a special role in my walk around Cape Cod. Its beacon was the first visible landmark in the pre-dawn stroll which began the trip. Returning to the light became a symbol of completing the trip. Finding out what was so special about this and other lighthouses became an important part of my effort to learn about Cape Cod.

Like Thoreau, Ralph Waldo Emerson visited Cape Cod. Decidedly unexcited, he wrote only a half a page in his journals about the Cape under the date of Sept. 5, 1853. The entry starts:

> Went to Yarmouth Sunday 5; to Orleans Monday 6th; to Nauset Light on the back side of Cape Cod. Collins, the keeper, told us he found obstinate resistance on Cape Cod to the project of building a light house on this coast, as it would injure the wrecking business. He had to go to Boston, & obtain the strong recommendation of the Port Society.

When the light was needed, the people of Cape Cod—or at least some of them—did not want it. Now that it is no longer needed, the people want it preserved forever. Therein lies a tale of mooncussers, shipwreckers, and beachcombers, three distinctly different professions of old.

Mooncussing was the practice of actively luring ships onto the shore so that they would be demolished and the cargo could be seized. Elizabeth Reynard wrote in The Narrow Land:

> At haggard sea corners of old Cape Cod men held lanterns high in the black nights of the seventeenth and eighteenth centuries. They swung the discs in a wide arc as though directing the pilotless ships over the Nauset Sea. Many a hemp-and-salt shipmaster mistook these swaying signals for the mastlights of other craft, turned to follow them, and ran on hidden bars... 'Mooncussers' was the name bestowed on these human harpies, who fed on the spoils of such moon-less disaster, who filched a lucrative plunder from the unchartable shifting of Race Point, Nauset, and Monomoy.

Other sources describe oxen being led along the shore with lanterns mounted on them to create the same impression.

Only one documented story of this deceptive practice can be found in the annals of Cape Cod. In Chatham several youths once went out to the outer beach and waved such lanterns, not to attract a ship, but as a practical joke to lure out the town's shipwreckers, who immediately entered their boats and hurried to the scene to partake in the spoils of what they perceived as a shipwreck.

Did mooncussing actually exist on Cape Cod? It is doubtful. If it did the perpetrators did a remarkably good job of keeping the act to themselves. Cape Codders have never been particularly close-mouthed when it comes to describing their own activities, however inappropriate. Henry C. Kittredge and other noted Cape historians get downright moralistic in denying the existence of such a dastardly practice, although Kittredge did not hesitate to capitalize on the appeal of the word by writing a book, The Mooncussers of Cape Cod. Kittredge states therein, "The mooncusser never existed on the sandy peninsula of the Cape, the wrecker and even the beachcomber have blithely appropriated his name," as, in fact, did Kittredge.

According to Kittredge, every town on the outer Cape had well organized shipwreckers, however, who raced to the site of each wreck to seize as much as they could of the surviving cargo, and of the wreck itself, for their own purposes. In one story an entire church emptied, preacher and all, when a wreck was reported. Another source states that when the wreckers from two towns arrived at the same shipwreck simultaneously, a split would be negotiated. Often seaweed or other worthless items were substituted for the real barrels of cargo when ship owners foolishly paid local people to salvage the contents—hoping, too often falsely, to buy "protection" against shipwrecking.

Why Kittredge and others are so moralistic in denying the existence of mooncussing while proudly describing the antics of the shipwreckers is somewhat intriguing. Although shipwrecking might have been less murderous than mooncussing, as shipwrecking was only done to vessels that came aground on their own accord, the fact remains that salvaged cargo belonged to the ship owners, not the salvagers. The two practices were equally illegal.

The story of Nauset Light's construction is also one of corruption, or at least of gross inefficiency. Stephen Plesenton, who oversaw construction of all lighthouses for the federal government for more than 30 years, had a strange and inexplicably close relationship with Winslow Lewis. Lewis was an unemployed former Massachusetts sea captain who 1.) prepared the description and specifications for the three lighthouses, 2.) underbid others for the contract to build the structures, and 3.) provided the original three lights at Nauset under his own patent for an outmoded type of light inferior to others available at the time. When his own nephew, Isaiah W. P. Lewis, was appointed later to audit the status of several lights, he cited Nauset as the type of fraudulent project conducted by contractors and recommended that one revolving light be substituted. Poor design and construction resulted in a need to replace the three lighthouses far sooner than necessary.

The original reason for building three lights was to distinguish Nauset from the one light at Race Point in Provincetown and the two lights in Chatham. The technology was then available for differing color lanterns and blinking patterns, but if it had been used, Lewis would not have gotten to sell three of his lanterns and to build three lighthouses, instead of one. Indeed, J. Brian West's *Life on the Edge* states that "each lantern had ten of Winslow Lewis' lamps, with thirteen-and-a-half inch reflectors, thirty in all!"

In fact, the current Nauset Light was originally not at Nauset at all, but one of the two positioned at Chatham. After the original set of three lighthouses at Nauset fell over the eroding cliff and their replacements (the "Three Sisters" currently on display in another location) also had to be moved, one of the original Chatham pair was transported north. Unlike the much witnessed trip down Nauset Road, no one is even sure how today's Nauset Lighthouse was moved from Chatham to Eastham.

Why do lighthouses fascinate us? Is it their illumination? Eleanor Roosevelt stated that she would rather light a candle than curse the darkness. Does their tall, erect posture provide a sense of strength, hope, and security? Or do they just enrich us by attracting tourists?

I asked Hawkins Conrad, chairman of the Nauset Light Preservation Society. His answer: "Different strokes for different folks." Some people come from far away to add another lighthouse to their lifetime list, just as they do with birds. For others, the lighthouse played a role in their past—some "smooched" under it, some swam under it, and some just walked by. For all, the lighthouse is a symbol of hopefulness and stability, as well as of Cape Cod itself. He concluded, "When I get to Cape Cod, I inevitably get sleepy, removed from the cares of the world."

If lighthouses help the process, so be it.

On a beautiful summer Friday evening at the end of my walk around the Cape, Sue and I walked the beach toward Nauset Light. It was a typical Friday evening on Cape Cod. As the sun went down, many people remained on the beach. Some, recently arrived for the weekend, showed vestiges of city clothing in their haste to get to the water before dark. They were obviously very happy to be there.

As night fell, I could not help but think of the contrasts between the first walk of my journey around Cape Cod and this one: sunrise versus sunset, Thoreau's lonely beach versus this day's crowded beach.

The Cape's shores were and still are attractive and wonderful places to be. Above all, they are happy place to be. Middle-aged couples hold hands. Children romp, splash, and build sandcastles, while parents sunbathe with watchful eyes before posing their families for photographs. Strangers and friends, each somehow shucked of their urban inhibitions, exchange warm greetings.

People, visitors and residents alike, come to Cape Cod because of the water. It mesmerizes and relaxes them. John F. Kennedy once said, "I always come back to the Cape and walk the beach when I have a tough decision to make. The Cape is the one place I can think and be alone."

Recognizing the lure of its waterfront, the City of Boston expects to complete a 43-mile "Harborwalk" from Dorchester to East Boston in about 10 years. The task has not been easy; some owners have cooperated, some have not. But, all new development plans must provide public access to the water. Linda Haar, director of Planning and Zoning for the Boston Redevelopment Authority, says, "I believe strongly in the idea of seamlessness. A seamless Harborwalk sends a signal to the city that there are parts of the waterfront which you can enjoy no matter how rich or poor you are." Her philosophy is even more relevant to the Cape.

Not everything is comparatively wonderful about Boston. "L Street Brownies," to the contrary, not many people want to go swimming in any of Boston's 43 miles. But, hard as it may be to believe that the mistakes made in the 20th century development of Cape Cod are more intractable than those made in the 18th and 19th century development of Boston, a seamless shore-walk is probably too much to hope for on Cape Cod.

We must recognize, however, that our shoreline is a public asset—the foundation of our economy and the heart of our being here. We should stop taking its beauty—and its utility—for granted.

Other Books from On Cape Publications

In the Footsteps of Thoreau:
25 Historic & Nature Walks on Cape Cod
by Adam Gamble

A Guide to Nature on Cape Cod & the Islands
edited by Greg O'Brien
$17.95

Cape Cod, Martha's Vineyard & Nantucket, the Geologic Story
by Robert N. Oldale
$14.95

Quabbin:
A History & Explorer's Guide
by Michael Tougias
$18.95

The Blizzard of '78
by Michael Tougias
$14.95

Baseball by the Beach:
A History of America's Pastime on Cape Cod
by Christopher Price

Sea Stories of Cape Cod and the Islands
by Admont G. Clark
$39.95

Haunted Cape Cod & the Islands
by Mark Jasper
$14.95

The Cape Cod Christmas Cookbook
by Mark Jasper
illustrated by Holly Shaker

www.oncapepublications.com